Keywords—Welding, health, cancer, disease, exposure, fumes, gases, literature review, noise, radiation, toxicology

Effects of Welding on Health—VI

Research performed by Tracor Jitco, Rockville, Maryland, under contract with the American Welding Society and supported by industry contributions.

An updated (July 1984-December 1985) literature survey and evaluation of the data recorded since the publication of the first report (1979). This series of reports is intended to aid in the understanding of the health effects of welding.

Performed by:

Winifred Palmer

May, 1987

Abstract

This literature review with 172 citations has been prepared for the Safety and Health Committee of the American Welding Society to provide an assessment of current knowledge of the effects of welding on health, as well as to aid in the formation of research projects in this area, as part of an ongoing program sponsored by the Society. Previous work has included studies of the fumes, gases, radiation, and noise generated during various arc welding processes. Referenced materials are available from Tracor Jitco.

Prepared for
Safety and Health Committee
American Welding Society
550 N.W. LeJeune Road, P.O. Box 351040
Miami, FL 33135

International Standard Book Number: 0-87171-298-9

American Welding Society,
550 N.W. LeJeune Road, P. O. Box 351040, Miami, Florida 33135

Personnel

AWS Safety and Health Research Committee

J.F. Hinrichs, Chairman	A.O. Smith Corporation
K.L. Brown, Vice Chairman	Lincoln Electric Company
M.E. Kennebeck, Jr., Secretary	American Welding Society
M.V. Anticoli	General Motors Corporation
*W.T. Delong**	Teledyne-McKay
*W.S. Howes**	NEMA
*E. Mastromatteo**	Consultant
Dr. M.T. Neu	Caterpillar Tractor Company
R.J. Simonton	INCO Alloys International
D.W. Walsh	California Polytechnic State University

**Advisor*

Foreword

(This Foreword is not a part of *Effects of Welding on Health VI*, but is included for informational purposes only.)

This literature review was prepared for the Safety and Health Committee of the American Welding Society to provide an assessment of current information concerning the effects of welding on health, as well as to aid in the formulation and design of research projects in this area, as part of an ongoing program sponsored by the Committee. Previous work included studies of the fumes, gases, radiation, and noise generated during various forms of arc welding. Conclusions based on this review and recommendations for further research are presented in the introductory portions of the report. Section 1 summarizes recent studies of the occupational exposures, while Section 2 contains information related to the human health effects of exposure to byproducts of welding operations. Section 3 discusses studies of the effects of welding emissions on laboratory animals and in vitro cell systems. Reference materials are available from Tracor Jitco, Inc.

Acknowledgments

Funds for this project were provided by the American Welding Society.

The American Welding Society gratefully acknowledges the financial support of the program by industry contributions.

Supporting Organizations

Air Products and Chemicals, Incorporated
Airco Welding Products
Akkus-Chalmers
Alloy Rods Division, The Chemetron Corporation
AWS Detroit Section
AWS New Orleans Section
Arcos Corporation
The Binkley Company
Catepillar Tractor Company
Chicago Bridge and Iron Company
Grove Manufacturing Company, Division of Kidde, Incorporated
General Electric Company
The Heil Company
Hobart Brothers Company
INCO Alloys International
Lincoln Electric Company
Miller Electric Manufacturing Company
National-Standard Company
A.O. Smith Corporation
Teledyne-McKay, Incorporated
Trinity Industries, Incorporated
Truck Trailer Manufacturers Association
Walker Stainless Equipment Company
Weld Tooling Corporation

Many other organizations have made contributions to support the ongoing program from May 1979 to the present.

Table of Contents

Section Three—Toxicologic Investigations in Animals and in Cell Cultures

List of Tables

List of Figures

Introduction

The health of workers in the welding environment is a major concern of the American Welding Society. To stay abreast of this subject, the health literature is periodically reviewed and published in the report Effects of Welding on Health. Five volumes have been published to date; the first covered data published before 1978, while the latter four covered time periods between 1978 and June, 1984. The current report includes information published between July, 1984 and December, 1985. It should be read in conjunction with the previous volumes for a comprehensive treatment of the literature on the Effects of Welding on Health.

Included in this volume are studies of the characteristics of welding emissions that may have an impact on the control technologies necessary to protect the welder (Section 1). In keeping with previous volumes, the health studies are organized according to the affected organ system. The respiratory tract, the primary route of exposure to welding emissions, is also a major target organ of a number of components of these emissions. Acute (e.g., metal fume fever, cadmium poisoning) as well as potential chronic respiratory effects (e.g.,emphysema, cancer) of welding emissions are of concern. However, chronic effects are not as well defined or understood and whether there is an excess risk of cancer from these exposures has not been established. Continued research in the form of epidemiologic studies, investigations with laboratory animals, and in vitro genotoxicity studies will help to resolve this question.

Executive Summary

Research on the health effects of welding continues to focus on the effects of chronic exposures to welding fumes on the respiratory tract. A problem inherent in much of this research is the difficulty in conducting studies on homogeneous populations due to the variability in welding processes and in working conditions. With time, this problem has become better understood, and today many investigators are attempting to relate their research results to exposures during specific welding processes.

The Respiratory Tract

Abnormal shadows are often seen in chest X-rays of welders. These shadows represent deposits of particles from welding fumes in the lungs (referred to as *arc welders' pneumoconiosis*). Normally these shadows are not associated with loss of lung function or diseases of the respiratory tract and, in some cases, they disappear after affected welders are removed from further exposure to the fumes.

Several studies performed during this report period indicate that welders exhibit a reduction in the volume of air that can be inhaled or expelled from the lungs (as measured by lung function tests). Other studies found no changes in lung capacity and attributed the absence of effects on lung function among welders so good ventilation in the work area or to the use of gas tungsten arc welding (GTAW) which generates little fume compared with most other commonly used welding methods.

The association of welding with bronchitis remains unclear. An excess of bronchitis was observed in welders in nine studies published during this report period. In three studies, the incidence of bronchitis could not be separated from the use of tobacco. One investigator suggested that smoking and welding may act synergistically in the induction of bronchitis. Bronchitis was observed more frequently in shielded metal arc welders than in gas metal arc welders, which supports the supposition that bronchitis may be related to the intensity of welding exposure.

The question of whether welders have an increased risk of lung cancer remains unresolved. In the past, results of cancer epidemiology studies of welders have been inconsistent, with several studies suggesting that the incidence of lung cancer may be elevated among welders. Nickel and hexavalent chromium, potential human carcinogens, may be present in significant quantities in stainless steel welding fumes. Past studies neither refuted nor supported the hypothesis that the lung cancer risk is elevated among stainless steel welders. Investigations currently in progress are focusing on populations exposed to welding fumes with known or suspected carcinogens. The results of these studies should begin to provide a much needed answer to the question of whether cancer in welders is associated with exposures to specific welding processes.

Severe acute respiratory distress can result from highly toxic chemicals, such as phosgene, that arise from interactions of degreasing agents such as 1,1,1-trichloroethane and ultra-violet light. Because of this, vapors from degreasing agents or paints can present a hazard in welding shops. Two incidents were described in which welders may have suffered respiratory distress from gas metal arc welding (GMAW) in an area where degreasing agents were used.

Biological Monitoring of Exposure to Welding Emissions

Industrial hygiene measurements of airborne concentrations of contaminants are exceedingly important for controlling exposures to welding emissions. However, they do not take into account variations in physiology and personal habits among workers and thus, do not reflect the amount of material actually taken up by the body from the work environment. In conjunction with determination of airborne exposure levels, biological monitoring, or the measurement of chemicals or their metabolites in the body fluids, may provide a means for estimating the actual dose of contaminants taken up by the body.

As in past years, considerable effort was expended during the current report period to identify substances that can be monitored in this way. The feasibility of applying biological monitoring techniques to elements such as chromium, nickel, lead, cadmium and manganese was examined. Blood and urine chromium levels, but not nickel levels, were found to be useful for estimating exposure to fumes generated by welding of stainless steel.

Technical Summary

The Exposure

Fumes

The concentration of solids in welding aerosols varies with the welding process, electrode, base metal, current, voltage, and base metal coatings. Shielded metal arc welding (SMAW) and flux cored arc welding (FCAW) produce more fumes, while GTAW produces much less fumes than other welding and allied processes. Because of its potential carcinogenicity to humans, the concentration of hexavalent chromium in fumes is important and has received a great deal of attention. The ratio of hexavalent to total chromium is much higher in fumes generated by SMAW and FCAW than in those generated by GMAW and GTAW.

Standardized methods for fume sample collection in laboratory settings and in the work place are important for providing realistic appraisals of risk and enabling comparison of results between laboratories. New standards for collecting samples of airborne particulates in the breathing zone and work area (Ref. 6) and for determination of fume generation rates and total fume emissions (Ref. 5) have been published by the American Welding Society.

The Swedish fume box for measuring fume emission rates (FER) generated by SMAW electrodes was modified by the British Welding Institute to enable determination of FER's from other welding processes (Refs. 94 and 99). FER's of carbon arc gouging, FCAW, GMAW, and GTAW, determined by the modified fume box, were generally within ranges found by other techniques (Ref. 94). Based on data obtained from fume box determinations, the British Welding Institute has established a computerized data base for storage and retrieval of data applicable to emissions from all arc welding processes (Ref. 100).

Using an aerosol photometer to obtain rapid measurements of welding fume concentrations at one second intervals, Glinsmann and Rosenthal (Ref. 46) showed that fume generation during welding is not uniform and continuous but rather there are wide and instantaneous fluctuations in fume concentration. Such fluctuations did not occur during oxyacetylene cutting.

The chemical composition of fumes may vary with fume generation rate as well as with the formulation of the electrode. Tandon et al. (Ref. 151) showed that the iron content of the fumes was directly proportional, and the fluoride content was inversely proportional, to fume generation rates. A strong correlation between the ratio of water-soluble hexavalent chromium to total chromium in the fume and the concentration of sodium and potassium in the flux was observed.

Substitution of sodium silicate for potassium silicate in basic electrode coatings reduced fume production by twenty five percent (Ref. 86). Limpel et al. (Ref. 81) found that the quantity of water-soluble, but not total fluorine, was considerably greater in fumes generated by electrodes containing potassium silicate than in those with sodium silicate. Other experiments showed that quantities of hydrogen fluoride (HF) and silicon tetrafluoride (SiF4) increased with the concentration of silica (SiO2) in the electrode coating, but only HF increased with the moisture content (Ref. 165). The levels of hexavalent chromium in the fumes were directly dependent on the quantities of sodium silicate, potassium and calcium levels in the electrode coating (Ref. 86).

Thorne and Hewitt (Ref. 157) showed that fume formation during brazing Ag-Zn-Cd-Cu alloys is related to the amount of bubbling in molten brazing flux. They concluded that exposures to cadmium fumes can be reduced by raising the pH of the fluoroborate flux to reduce hydrogen gas formation and by lowering the concentrations of compounds (boric acid and potassium hydrogen fluoride) in the flux that are largely responsible for bubble formation.

The issue of which technique is best for measuring hexavalent and total chromium in fumes remains controversial. Dare et al. (Ref. 29) criticized the interlaboratory round robin validation study of the Blakely and Zatka method for hexavalent chromium determination (Ref. 22) for not taking into account the short-lived hexavalent chromium species detected in fresh GMAW fumes collected by water impingement (Refs. 53 and 156). Zatka (Ref. 166) suggested that this hexavalent chromium species derives from a chemical reaction within the impingement collector fluid rather than by reactions within the fume solids. Zatka also modified his initial method (Ref. 22) for hexavalent chromium determination to suppress the small amount of hexavalent chromium that can form during digestion of trivalent chromium in hot alkaline solutions.

Gases

Ozone, carbon monoxide, carbon dioxide and nitrogen oxides are the principal gases generated by welding. The quantities of nitrogen oxides, carbon monoxide, carbon dioxide and methane released during SMAW varies considerably with the electrode (Ref. 158). Nemcova (Refs. 108 and 109) reported finding high concentrations of nitrogen oxides and ozone during plasma arc cutting of steel, aluminum, and copper. Nitrogen oxide levels were relatively high during argon-shielded GTAW of copper and aluminum.

Radiation

Eriksen (Ref. 37) demonstrated that a short burst of high levels of ultraviolet (UV) radiation occurs during the initial phase of arc ignition during GMAW of aluminum. The intensity of this UV overshoot was more than ten times that of the UV light emitted when the arc was burning "smoothly". Because of the intensity of the UV overshoot, the unprotected eye at a distance of 0.5 meters may suffer welder's flash after exposure to radiation from only one ignition.

Production Coatings

Organic compounds, metal oxides, and toxic gases may be released when welding metals coated with paints or primers. Moreton (Ref. 98) and McMillan (Ref. 91) reviewed the hazards of welding coated or contaminated surfaces. Moreton described and evaluated standard procedures recommended by representatives of the welding and paint industries in 1964 and 1968 for assessing the toxicity of fumes released from flame cutting or welding of primed metals. The need for new standards which would allow realistic appraisal of the hazards associated with welding and cutting of primed materials was stressed (Ref. 97).

Degreasing Agents

Fumes from degreasing agents or paints can present a major hazard in welding (Refs. 9 and 91). Chlorinated hydrocarbons such as trichloroethylene, perchloroethylene and 1,1,1-trichloroethane can decompose in the presence of ultraviolet radiation into highly toxic compounds such as phosgene and dichloroacetyl chloride.

Two incidents were described in which photochemical decomposition products resulting from reaction of degreasing agents with UV light produced by GMAW may have been the cause of significant respiratory effects. In the first, eleven welders complained of coughing, breathlessness, chest tightness, and irritation of the throat and eyes while performing GMAW, but not SMAW. A degreasing operation was conducted on the shop floor in the vicinity of the welders where there was good general ventilation. Ambient concentrations of trichloroethylene and trichloroethane in the work environment were too low to have caused noticeable health effects and phosgene could not be detected. Removal of the GMAW operation to an area distant from sources of chlorinated hydrocarbons successfully resolved the problem (Ref. 135).

In the second incident, a Swedish welder developed pulmonary edema after performing GMAW of mild steel in an environment containing high levels of 1,1,1-trichloroethane. Although 0.4 ppm ozone was detected when welding under the same conditions in a glove box, insufficient concentrations of toxic gases were detected when welding in an open room similar to the one used by the affected welder to have accounted for the pulmonary edema. (Ref. 56).

Noise

Impulse noise is more harmful to hearing than continuous steady state noise of the same energy level. Tests performed at a shipyard plate welding workshop indicated that grinding and carbon arc work do not produce impulse noise, while welding (especially GMAW) had the highest impulse noise levels (Ref. 78).

Effects of Welding on Human Health

Respiratory Tract

The physiological consequences of pneumoconiosis ["the accumulation of dust in the lungs and the tissue reaction to its presence", (Ref. 64)] are related to the fibrogenicity and amount of deposited dust. Several investigators studied the extent of fibrosis in the lungs of subjects with arc welders' pneumoconiosis. Most found that scar tissue formation was minimal, that fibrosis rarely progressed into a disabling condition, and opacities seen in chest X-rays of those with pneumoconiosis sometimes regressed with time after cessation of welding exposures. Such results indicate that arc welders' pneumoconiosis is, for the most part, a benign condition (Refs. 130, 131, 137, 142, 144, and 147).

The effects of welding on pulmonary function were examined in twelve studies. Half of these found no substantial abnormalities (Refs. 85, 139, 142, 171, 130, and 136). In two studies (Refs. 136 and 139), the absence of effects on pulmonary function among welders was attributed to good ventilation in the work area. A third study (Ref. 171) attributed the overall good health status and lack of pulmonary function abnormalities to the use of GTAW which generates little fume compared with most other commonly used welding methods.

An age-related reduction in pulmonary function among welders was found by Zober and Weltle (Ref. 172). Mao (Ref. 88) reported an association between pneumoconiosis and abnormal pulmonary function tests, but three other investigators found no such correlation (Refs. 85, 142 and 172). Mur et al. (Refs. 105 and 106) reported that shielded metal arc welders had significant reductions in pulmonary function compared with gas metal arc welders. In some cases, welders who worked in confined spaces had reduced lung function compared with those in well ventilated areas. Schneider et al. (Ref. 137) noted a reduction in pulmonary function which appeared to be related to the duration of welding exposure.

An association between exposure to welding emissions and bronchial hyperreactivity was suggested by Barnhart et al. (Ref. 15), who observed a decrease in pulmonary function in welders only after challenge with bronchodilators. Inconsistencies between studies may be related, in part, to the inability of pulmonary function tests to routinely detect early changes in the lungs. In general, this research suggests that welding may reduce selected parameters of pulmonary function and that fume emission rates and ventilation in the work area are influential factors.

An elevated incidence of bronchitis was found in welders in nine studies. Zober et al. (Refs. 171 and 172) attributed bronchitis solely to tobacco use. A substantially elevated frequency of bronchitis among welders was also reported by Mal'ik et al. (Ref. 85). However, in that report, no attempt was made to account for the high rate of tobacco use among the cohort. That smoking and welding may act synergistically in the induction of bronchitis and reduction of pulmonary function was suggested by Schneider et al. (Ref. 137) and Mur et al. (Refs. 105 and 106), respectively. Mur et al. (Refs. 105 and 106) observed bronchitis more frequently in shielded metal arc welders than in gas metal arc welders, which implies that bronchitis may be related to the type and extent of exposure. Whether or not welding alone can produce bronchitis remains unclear.

Cancer

Langard and Stern (Ref. 80) reviewed twenty one cancer epidemiology studies of welders published in the international literature before 1984. Of these, five studies showed a significantly elevated cancer risk among welders, and only one examined a cohort of stainless steel welders. Stainless steel welding fumes contain hexavalent chromium and nickel compounds which may be carcinogenic. Because of this, it has been suggested that epidemiologic studies focus on stainless steel welders (Ref. 80).

Two epidemiologic studies of nickel-and chromium-exposed welders were published in 1985. Using a case-referent approach, Gerin et al. (Ref. 45) showed that persons with excessive nickel and chromium exposures exhibited a threefold excess of lung cancer while there was no statistically significant association between nickel exposure and the risk of cancer development in other organs. A retrospective epidemiologic study of the cancer risk from exposure to nickel and chromium in 1221 welders employed in 25 German factories was conducted by Becker et al. (Ref. 18). They found that the cancer mortality rate was significantly increased in welders, but the number of cancer cases was too small to enable calculation of cancer rates in specific organs.

Negative results were obtained in three cancer mortality studies. Milham (Ref. 92) determined the number of deaths from cancer among the death records of 486,000 adult men filed in the State of Washington between 1950 and 1982. A significant increase in the cancer incidence was not observed among the welders. Newhouse et al. (Ref. 110) found no significant increases in the mortality rates from cancer among 1027 welders who worked in a British shipyard between 1940 and 1968. Similarly, Esnault et al. (Ref. 38) found no significant excess of deaths from cancer among 100 welders who worked at a French shipyard for a period of 16 years. A possible combined effect of exposure to polycyclic aromatic hydrocarbons and welding emissions was found by the United Automobile Workers Union. A two- to five-fold excess in the number of deaths from cancer of the lung, digestive organs, testes and leukemia was observed in millwrights and welders exposed to fumes from welding and coal tars (Ref. 138).

Mastromatteo (Ref. 90) reviewed the epidemiologic studies of nickel published since 1977 and concluded that there was insufficient evidence to associate nickel with respiratory cancer in nickel welders. In his review of published epidemiologic and animal studies of the carcinogenicity of iron oxide, Stokinger (Ref. 147) concluded that iron oxides are not carcinogenic.

Metal Fume Fever

Metal fume fever is caused by excessive inhalation of metal oxide fumes. Fever, chills, general malaise, joint pains, cough, sore throat, chest tightness and fatigue usually appear four to twelve hours after exposure and last from one to two days. Diagnosis of metal fume fever is often difficult because its symptoms resemble those of a number of upper respiratory tract illnesses. Barnhart and Rosenstock (Ref. 14) reported a case of a welder with cadmium poisoning that was misdiagnosed as metal fume fever. Because of the potential seriousness of cadmium poisoning, they emphasized that possible exposure to cadmium fumes should be considered whenever patients show symptoms of metal fume fever.

Effects on the Ear and Hearing

Impulse noise may be more damaging to hearing than continuous noise of the same energy level. Platers and welders exposed to impulse noise developed hearing loss with shorter exposures than did workers in a cable factory exposed to continuous noise (Ref. 87).

Effects on the Eye and Vision

Photokeratitis (welder's flash) is a painful inflammation of the cornea which results from exposure to ultraviolet radiation. Symptoms do not appear until several hours after exposure. Millodot and Earlam (Ref. 93) found the cornea to be less sensitive to touch during the first several hours following exposure to radiation from the welding arc.

Pterygia, membranous growths which extend across the outer eye from the conjunctiva to the cornea, are thought to be caused by ultraviolet radiation. Karai and Horiguchi (Ref. 71) found a significant relationship between the duration of employment as a welder and the incidence of this lesion.

Two studies suggested that exposure to the welding arc accelerates changes in the eye which normally occur with aging. The first found changes in the number of cells in the corneal endothelium. These changes have no known effect on corneal function (Ref. 72). The second study indicated that welding may accelerate the formation of senile cataracts (Ref. 84). Conversely, Dvorak et al. (Ref. 34) found no changes in the lens that could be attributed to shielded metal arc welding.

The time required for restoration of visual acuity following exposure to strong light was reported by Dobromyslov (Ref. 31) to be significantly decreased in gas metal arc welders whereas no such changes were observed by Gos et al. (Ref. 52). The latter investigators noted that welders were better able to differentiate shapes at minimum illumination than were controls.

Sensitivity to Fume Components

Four cases of severe contact dermatitis, with itchy rashes on the ankles, hands and forearms, severely fissured lip lesions, bloody nasal discharge, muscle and joint pain, and inflammation of the esophagus, were related to the frequent and intensive use of high-chrome welding rods (Ref. 102).

Effects on the Musculoskeletal System

Herberts et al. (Ref. 60) showed that welders had a significantly greater incidence of shoulder tendonitis than did controls. Both Herberts et al. (Ref. 60) and Stone (Ref. 148) reported that static work may entail a higher risk for chronic shoulder pain and that the most common repetitive strain injury among welders is caused by static load strain.

Effects on the Urogenital Tract

Cadmium can cause kidney damage which is first manifested by urinary excretion of serum proteins (proteinuria). Chiesura et al. (Ref. 24) found that blood and urine cadmium levels in 16 cadmium-exposed persons were related to the duration and severity of exposure. One case of nephropathy was attributed to cadmium exposure.

Elinder et al. (Ref. 35 and 36) characterized the nature of renal effects in 60 cadmium-exposed workers. Urinary beta-2-microglobulin concentrations correlated well with levels of urinary cadmium and with the estimated cumulative cadmium dose. Cadmium-exposed workers with a history of kidney stones had significantly higher urinary cadmium levels and tended to have higher urinary beta-2-microglobulin levels than those with no history of stones. The low molecular weight protein, urinary beta-2-microglobulin, remained elevated in most subjects even six years after exposure had ceased, indicating that cadmium may cause irreversible or chronic kidney damage.

During normal kidney function, low molecular weight proteins, and only low levels of high molecular weight proteins, are filtered through the glomeruli and resorbed through the renal tubules. Thus, the presence of low molecular weight proteins in the urine is indicative of tubular dysfunction while high molecular weight urinary proteins may be indicative of glomerular dysfunction. Falck et al. (Ref. 40) observed both high and low molecular weight proteins in urine samples from seven cadmium-exposed workers and concluded that cadmium causes glomerular and tubular damage. Elinder et al. (Ref. 36) found that levels of low molecular weight proteins were substantially greater than levels of high molecular weight proteins. They argued that decreased tubular resorption, rather than increased glomerular permeability for larger proteins, was responsible for the changes.

Effects on the Emdocrine System

Smirnov et al. (Ref. 141) examined urinary hormone levels in welders and persons in other occupational groups as a measure of strain and physical stress. Adrenalin and noradrenalin levels were significantly higher in electric arc welders working in poor conditions than in welders with generally satisfactory work conditions.

Effects on the Teeth and Oral Cavity

A survey of Swedish commercial divers indicated that thirty six of sixty six who performed underwater electric arc welding sensed a metallic taste in their mouth, possibly derived from degeneration of dental amalgams, when working with electrical equipment under water (Ref. 122). The appearance of dental amalgam restorations of twenty nine divers who performed underwater welding or cutting differed significantly from those of eleven divers who had not worked with underwater electrical equipment.

Effects of Hyperbaric Pressure

Bjorseth et al. (Refs. 20 and 21) discussed the need for research into the toxic effects of gases and fumes at hyperbaric pressures representative of the great oceanic depths experienced by underwater welders employed in off-shore oil operations. Preliminary work indicated that hyperbaric pressure can cause changes in the chemical composition and emission rates of welding fumes and gases. The toxic effects of chemicals may change under high hydrostatic pressure and chemical exposures may alter the physiological stress already inherent in deep sea diving.

Adverse physiological effects, including nervous excitation and changes in pulmonary function, can result from exposure to hyperbaric conditions in the absence of chemical exposures. That the stress already imposed upon the body by hyperbaric pressure can alter the effects of chemicals was suggested in studies of a few medications, anesthetics, and breathing gases which showed that effects seen at normobaric pressures cannot be extrapolated to hyperbaric conditions. New research into the effects of exposure to gases and fumes under hyperbaric conditions is necessary for the development of exposure guidelines appropriate for use in underwater welding.

Biological Monitoring

Investigations of urine or blood levels, or both, of nickel and chromium indicate that chromium, but not nickel, can be useful for monitoring worker exposure (Refs. 2, 168, and 170). Akesson and Skerfving (Ref. 4) found no correlation between urinary nickel concentrations and the extent of nickel exposure.

Zober (Refs. 168 and 170) determined concentrations of nickel and chromium in the breathing zone and body fluids of twenty stainless steel welders. A linear relationship existed between exposure levels and post-shift concentrations in the urine and plasma for both hexavalent and total chromium. Preshift urinary chromium levels increased throughout the week indicating an accumulation of chromium in the body. It was concluded that workplace exposures to chromium can be successfully monitored by either urine or plasma analysis.

Koshi et al. (Refs. 75 and 76) measured the frequency of sister chromatid exchanges, chromosomal aberrations, and the number of chromosomes in lymphocytes from stainless steel welders. No significant difference was seen in the sister chromatid exchange frequency between welders and controls. Chromosomal aberrations, including chromosome gaps, aberrant metaphases, and chromatid and chromosome gaps, occurred slightly, but significantly more frequently in welders than in controls.

Sjogren et al. (Ref. 140) examined the relationship between urinary aluminum levels and exposure to fumes generated by GMAW of aluminum. Their study indicated that aluminum inhaled from welding fumes may be retained for long periods of time. Part of the inhaled aluminum is excreted rapidly in the urine following exposure while the remainder may be stored in the body and excreted slowly.

Diagnostic tests which measure the effects of lead on the heme system, such as measurement of zinc erythrocyte protoporphyrin (ZEP) and urinary coproporphyrin, are generally considered to be useful indicators of the biological effects of absorption of lead by the body, but are less useful than blood lead levels for monitoring acute lead exposures (Refs. 17 and 61).

Williams (Ref. 164) reported the case of a welder who had a massive acute exposure to lead while performing a temporary assignment in a lead-acid battery factory. The welder's blood lead levels were highly elevated; lead levels of 240 and 300 g/dl were measured on two separate occasions within two weeks after exposure. Urinary lead and coproporphyrin concentrations were only slightly elevated whereas ZEP rose for eight weeks and then slowly decreased.

That ZEP may be a better biological screening parameter than blood lead levels for chronic lead exposure was reported by Kalnas and Alleyne (Ref. 70) who measured urinary aminolevulinic acid (ALAU), blood lead and ZEP in 142 lead-exposed workers. ZEP, but not blood lead levels, reflected symptoms of lead exposure in workers employed two years or less. ALAU levels tended to be indicative of longer term exposure.

Huel et al. (Ref. 62) examined whether cadmium and lead absorbed during occupational exposures are capable of reaching the human fetus by determining the concentrations of these metals in hair samples from newborns shortly after birth. The concentration of cadmium and lead in hair from exposed mothers and of cadmium in hair from their offspring were more than twice as high as concentrations in hair from unexposed controls. However, there was no correlation between maternal and newborn lead levels. This indicated that systemic cadmium, but not lead, exposure can be quantified by hair analysis of either the mother or the newborn.

Gorban et al. (Ref. 49) determined the manganese content in hair samples collected from 228 welders who performed CO2-shielded GMAW. The investigators concluded that a direct relationship exists between the extent of manganese exposure and the concentration of manganese in hair.

Toxicologic Investigations in Animals and Cell Cultures

Animal Studies

Lam et al. (Ref. 79) found that exposure of guinea pigs for six days to 5 mg/m3 zinc oxide, a concentration equivalent to the threshold limit value (TLV - Ref. 1), caused alterations in pulmonary function. The lung weight was markedly increased and did not return to normal during the three-day post-exposure examination period. On the basis of this study, they suggested that the TLV may not be low enough to protect exposed workers.

That exposures to mixtures may be more harmful than individual components was demonstrated by Sylvestre and P'an (Ref. 150) who studied the effects of thirty-day exposures to ozone, carbon monoxide (CO), nitrogen dioxide (NO2), and manganese dioxide (MnO2) particles on the mouse lung. All experimental groups had emphysema and dilated alveoli. Emphysematic lesions were less extensive in animals that received a mixture of particulates and gases than in those that received gases or particulates alone. However, the mixture of gases and particulates caused the most overall damage and produced numerous areas of inflammation, edema, and dilated blood vessels.

Pulmonary fibrosis was not observed in the lungs of rats treated by intratracheal instillation with fumes generated by CO2-shielded GMAW (Ref. 50). LD50's, determined by intraperitoneal injection, indicated that the acute toxicity of particulates decreased as the welding current increased and that fumes from high alloy steel were substantially more toxic than those from low alloy steel.

To investigate the fibrogenic potency of welding fumes, Weller and Reichel (Ref. 163) injected welding fumes containing 88.5 percent iron, 7.5 percent manganese, 3.2 percent silicon, and 0.9 percent aluminum into the rat peritoneal cavity. After three months, a large number of macrophages and connective tissue fibers were associated with dust deposits. The fibrosis did not progress during the remaining nine months of the study. The investigators concluded that the discrete fibrotic changes in the peritoneum parallel changes seen in the human lung and that welding dusts cause a limited fibrotic response which does not progress into massive fibrosis. In a similar experiment, Malik et al. (Ref. 85) observed no signs of fibrosis in guinea pigs injected intraperitoneally with particulates generated by welding.

Olah et al. (Ref. 118) examined the lungs of Wistar rats twelve weeks after intravenous injection of welding fumes collected from SMAW and argon-shielded GTAW of austentitic steel. SMAW fumes produced a mild to moderate fibrogenic response in the lungs, but GTAW fumes were not fibrogenic.

Kalliomaki et al. (Ref. 69) compared the retention and clearance rates in rats exposed by inhalation to fumes from GMAW and SMAW of stainless steel. Chromium generated by GMAW was cleared from the lungs much more slowly than chromium from SMAW. Nickel cleared rapidly at first and then, after several days, cleared much more slowly. A bimodal excretion pattern was noted for chromium. At first, chromium levels fell rapidly in the urine, with a half-life of eight hours. Two days after exposure ceased, urinary chromium excretion decreased to a half-life of thirty days.

In Vitro Studies

Bacterial Assays. Using a modified Salmonella/ Ames assay, Biggart (Ref. 19) found that both the gas phase and particulates from mild steel welding fumes contained mutagens. These findings differ markedly from previous reports and warrant follow-up studies.

Mammalian Cell Studies. Hansen and Stern (Refs. 57 and 58) studied the cytotoxicity and transforming effects of welding fumes and their components on cultured baby hamster kidney cells (BHK-21) and Syrian hamster embryo (SHE) cells. Hexavalent chromium, but not trivalent chromium compounds, were cytotoxic and transforming (see *Mammalian Cell Studies*). The transforming potency and toxicity of fumes from SMAW of stainless steel were ten times greater than those from GMAW of stainless steel. Fumes from SMAW and GMAW of mild steel were only very weakly toxic and did not cause cell transformation.

Potebnia et al. (Ref. 127) tested the toxicity and cocarcino-genicity of three welding fume samples in SA7 adenovirus infected cultured rat kidney cells and hamster embryo cells. All fume samples enhanced the adenovirus-induced oncogenic transformation of both cell lines. When transformed cells were injected into animals, tumors developed more rapidly from cells treated with welding fumes and virus than from those treated with virus alone.

Conclusions

Welding emits fumes and gases which may cause adverse health effects. Emissions vary widely with the process and can be controlled to some degree by choice of method, electrode and filler material. Industrial hygiene surveys indicate that, with the exception of GTAW, the TLV's for fumes and components of welding emissions are occasionally exceeded with commonly used welding methods. In many cases, these excess exposures result from the lack of adequate ventilation (Ref. 133). Source ventilation is the best method for reducing fume exposure, but even with local exhaust, some exposure to fumes and gases may occur. Sufficient information is currently available to warrant keeping exposures to a minimum. Training programs may be needed for management and welders alike, demonstrating the economic and health advantages of reducing exposures.

Unanswered questions concerning health issues persist, including the effects of welding on pulmonary function, the health status of the respiratory tract, cancer rates in welders, and the mutagenicity of welding fumes. These issues have been addressed in previous volumes of the Effects of Welding on Health (Refs. 123 and 161) and are only briefly iterated here. New issues that have been brought out by recent research reports concern the effects of impulse noise on hearing and the need for data on the combined effects of hyperbaric pressures and welding exposures.

Pulmonary Function

Chronic exposure to low concentrations or acute exposure to high concentrations of components of welding emissions—including ozone, nitrogen oxides, chromium and nickel compounds—may injure the respiratory tract. However, whether or not deficits in pulmonary function, or the development of respiratory diseases such as bronchitis and emphysema, are associated with welding exposures is not firmly established. The generation of pollutants that may injure the lung varies with welding methods and the metal welded. Therefore, studies of the incidence of respiratory diseases in welders should be carefully designed and study populations selected in a manner which will allow correlation of results with exposures to specific welding processes. In addition, a wide variety of pulmonary function tests are available. The applicability of these tests for general medical screening of welders should be examined.

Cancer

Fumes generated by welding stainless steel contain potentially carcinogenic chemicals, in particular hexavalent chromium and nickel compounds. Epidemiologic studies suggest that welders have an increased relative risk of thirty to forty percent for developing lung cancer (Ref. 145). However, this elevated risk has not been associated with exposure to emissions from specific welding procedures. Two studies, reported in this volume of *Effects of Welding on Health*, indicate that stainless steel welders have an elevated risk of cancer, but only one of these studies had sufficient data to point to the lung as a target organ. Continuation of these studies, and completion of other studies now in progress (Refs. 41 and 145), should begin to provide a much needed answer to the question of whether exposures to specific fumes account for the elevated cancer risk observed among some welding populations.

In Vitro Studies

Screening of chemicals for oncogenic transformation and mutagenicity in short-term biological assays is important for the identification of chemicals potentially carcinogenic or mutagenic to man. Information derived from short-term tests may become the basis for the longer-term bioassays of chronic toxicity in animals and for selection of populations for epidemiologic studies. Hexavalent chromium compounds cause mutations in bacteria, cultured mammalian cells and experimental animals. Nickel has produced inconsistent results in short-term mutagenicity assays; however, transformation of cultured cells has been accomplished with nickel compounds.

In the past, welding emissions from stainless steel, but not mild steel, have been mutagenic in short-term tests. A recent study (Ref. 19), using a modified

Salmonella/Ames test, indicated that gases and fumes from mild steel may be mutagenic. Because of the implications of these results, the modified test system should be examined and similar tests conducted by other investigators to determine whether or not there are previously unrecognized mutagens in welding fumes.

Two studies found chromosomal aberrations, including an increase in the number of chromosome gaps, in nickel-exposed workers (Refs. 75, 76, and 162). According to Aitio (Ref. 2), the importance of such gaps is not understood. The increase in chromosome gaps in nickel-exposed workers may be an area worthy of further research.

Combined Exposures

TLV's and other regulatory exposure levels are based on the toxicity of single elements or compounds. Combined or synergistic interactions are not considered, and the effects of mixtures may be quite different from the additive effects of individual components. For example, Sylvestre and P'an (Ref. 150) demonstrated that the effects on the mouse lung caused by inhalation of a mixture of gases (ozone, CO, and NO2) alone, or of MnO2 particles alone, were less severe than those produced by the combination of gases and MnO2. In some cases, mixtures may be less toxic than the additive effects of their individual components (Ref. 63). To help elucidate the health effects of welding emissions, laboratory studies of the cumulative effects of exposure to different components of welding fumes remain an important area of investigation.

In addition to mixed exposures to agents generated by welding, workers may experience adverse effects from combined exposures to welding emissions and pollutants from other sources in the workplace. An elevated incidence of cancer among welders with a combined exposure to polycyclic aromatic hydrocarbons and welding emissions was reported by Silverstein et al. (Ref. 138).

Fumes from degreasing agents or paints may present a major hazard in welding shops. Chlorinated hydrocarbons, such as trichloroethylene, perchloroethylene and 1,1,1-trichloroethane, can decompose in the presence of ultraviolet (UV) radiation into highly toxic compounds such as phosgene and dichloroacetyl chloride. Two incidents were described during this report period in which photochemical decomposition products resulting from reaction of degreasing agents with UV light produced by GMAW may have been the cause of respiratory problems (Refs. 56 and 135). In neither of these incidents could the agents responsible for respiratory distress be positively identified although the relationship between ultraviolet radiation generated by GMAW and chlorinated hydrocarbons seems certain. Studies of the chemical and toxicologic consequences of GMAW in the vicinity of chlorinated hydrocarbons are necessary to further understand the hazards present in such an environment.

Bjorseth et al. (Refs. 20 and 21) described the need for research into the combined toxic effects of hyperbaric pressures and welding emissions. Preliminary reports indicate that the composition of welding emissions are modified at high pressures and that the effects of chemicals on animals differ from those occurring at normal barometric pressures. Studies are needed to develop exposure guidelines to protect underwater welders in offshore oil exploration. According to Bjorseth, the assessment of short-term risk assessment is the most immediate need.

Effects on the Ear and Hearing

The Academy of Finland, Research Council for Technology, has been performing a major research project on occupational hearing. They found that impulse noise may be more harmful to hearing than continuous noise (Ref. 87). The sensitivity to certain sound frequencies may be more easily affected by impulse noise than by continuous noise. Future studies should address the adequacy of ear protectors to protect against hearing loss from exposure to impulse noise.

Effects of Welding on Health VI

Section One
The Exposure

Welding generates fumes, gases, and electromagnetic radiation with known adverse health effects. The composition of welding emissions varies substantially with the welding process. Of the most common welding processes, fume generation is lowest with gas tungsten arc welding (GTAW) and highest with shielded metal arc welding (SMAW) and flux-cored arc welding (FCAW). Gas metal arc welding (GMAW) of aluminum generates the highest ozone levels among widely used welding processes.

The degree of exposure to welding emissions can vary widely from workplace to workplace, as well as among the work areas within individual workplaces (Ref. 30). Differences in natural and mechanical ventilation systems account for much of this variation (Ref. 133). Local fume extraction or source ventilation is considerably more efficient in reducing exposure levels than general ventilation (Refs. 9, 27, 30, 77 and 114), but even with local ventilation techniques, airborne concentrations of fumes and gases in excess of threshold limit values (TLV's—Ref. 1) have been recorded (Ref. 30). While not always practical, exposures can also be controlled by using welding procedures and electrodes that generate lower quantities of emissions. Several industrial hygiene surveys demonstrated that exposure levels to welding fumes and gases vary with the welding process. Surveys performed by Zober et al. (Refs. 168 and 171) indicated that exposures to chromium, nickel, and total welding fume were low during GTAW and GMAW, but not during SMAW, of stainless steel. Grothe et al. (Ref. 54) examined over 700 breathing zone and ambient air measurements taken in German workplaces where SMAW, GTAW, and GMAW of stainless steel were performed. With the exception of SMAW and welding of metals with high nickel contents, concentrations of chromium and nickel in welding fumes were generally below maximum permissible levels.

Van der Wal (Ref. 160) surveyed breathing zone concentrations of particulates and gases generated during steel welding in Dutch factories. The Dutch occupational health standards [equivalent to the TLV's in the USA] were occasionally exceeded for total dust during SMAW and GMAW, but not GTAW. Total and hexavalent chromium exposures only exceeded the standards during SMAW of stainless steel. Of the measurements made for nitric oxide (NO), nitrogen dioxide (NO2), and ozone during GMAW, GTAW, and SMAW, only concentrations of ozone generated by GMAW of aluminum were above established Dutch standards.

1. Fumes

Welding fumes are composed of metal oxide particles which originate primarily from the filler metal and the electrode coat or core materials. Concentrations of fume constituents vary with the welding process, electrode, base metal and material that may be coated on the base metal. The welding parameters (voltage and current) may also influence the relative concentrations of individual components of welding fumes. For example, Olah et al. showed that the concentrations of chromium, nickel and iron generated by SMAW increased, but manganese remained relatively constant as the current was increased from 70 to 170 amps (Ref. 116 and 120).

1.1 Analysis of Welding Fumes. Standardized methods for collection of welding fume samples, both in laboratory settings and in the workplace, are important for providing realistic appraisal of risk and enabling comparison of results. A standard method for collecting samples of airborne particulates in the breathing zone and work area was developed by the American Welding Society (Ref. 6). The standard emphasizes that environmental conditions, electrode, welding parameters, base metal, surface contaminants, and design of the welding helmet should be carefully noted when samples are collected, as all of these factors can influence the exposure of the individual welder.

The American Welding Society also developed a standard laboratory method for determination of fume generation rate (FGR) and total fume emission. This standard presents a test chamber design that can be used for automatic, semiautomatic, or manual processes (Ref. 5).

Specifications for a fume box used to measure fume emission rates (FER) generated by SMAW electrodes were designated in the 1975 *Swedish Standard for Classification of Manual Metal Arc Electrodes into Fume Classes* (Ref. 11). The British Welding Institute modified this fume box to enable determination of FER's for other welding processes (Refs. 95 and 99). The feasibility of using the modified fume box to determine FER's during carbon arc gouging, FCAW, GMAW, and GTAW was examined (Ref. 95). In general, FER's were within ranges measured by other techniques. As expected, GTAW produced much less fume than other processes; FER's for GTAW and hot wire GTAW were about 30 to 100 and 4 to 10 times lower, respectively, than those for GMAW. Figure 1 shows the fume emission rates for various welding conditions calculated in terms of grams deposited metal or arc—on time (Ref. 99).

Based on data obtained from fume box determinations, the British Welding Institute established a computerized database for storage and retrieval of data applicable to emissions from all arc welding processes (Ref. 100).

Glinsmann and Rosenthal (Ref. 46) evaluated an aerosol photometer, which operated on the principle of light scattering, for monitoring welding fume levels in the workplace. The utility of the photometer was limited because its sensitivity varied with the chemical composition and size of the particle. Since particle sizes change rapidly as welding fume particles agglomerate after formation, it was not possible to develop accurate calibration factors for the equipment. However, immediate, onsite measurements could be obtained that might be useful for the rapid identification of areas with elevated fume concentrations as well as for rough assessments of the effectiveness of engineering controls.

Measurements taken at one minute intervals with the aerosol photometer revealed a smooth variation in fume concentrations during and after welding (Figure 2). With short-term welding operations in semi-enclosed and enclosed areas, fume concentrations increased for two to four minutes after the arc was struck; the decrease in fume levels after the termination of welding was slightly more rapid. When fume levels were measured at one second intervals (Figure 3), wide and instantaneous fluctuations in fume concentration were seen during arc welding. Such fluctuations did not occur during oxyacetylene cutting.

Based on the Baum and Mulholland theory for particle coagulation in buoyant plumes, Olander (Ref. 121) developed a model for calculating the particle number flow, particle number concentration, and the mass flow as functions of plume height. According to Olander, use of this model allows recommendations for suitable vertical temperature gradients in welding work areas and quantification of requirements for local exhaust systems.

Akbarkhanzadeh (Ref. 3) described a personal air sampler developed for use in an epidemiologic study of welding workplaces. The sampler was designed to operate continuously over an extended time and to collect both particulates and gases. The sampling train consisted of a sampling head, which housed the particulate filter and was placed inside the welder's helmet, followed by parallel assemblies for analysis of carbon monoxide and nitrogen oxides. The personal monitoring system could sample a total volume of 150 liters of air at a flow rate of about 300 ml/minute from SMAW of primed mild steel without a significant reduction in flow rate due to buildup on the particulate filter.

To investigate the effectiveness of welding helmets in reducing fume exposure, Goller and Paik (Ref. 47) compared total fume concentrations inside and outside the welding helmets. Personal air samples were collected from eight welders at four positions on the body (inside the helmet, right and left shoulder, and the front breast pocket) during inert gas-shielded FCAW and SMAW. The fume concentrations outside the helmet varied with the type of welding and the welder's posture. Breathing zone concentrations ranged from thirty six to seventy one percent of the fume concentrations measured outside the helmet. In response to this report, Jenkins and Moreton (Ref. 67) commented that the positioning of the sampling head within the helmet may have exposed the filter to moisture from exhaled breath which could have decreased its collection efficiency. They also stated that the filters used in this experiment may become clogged within about two hours during SMAW or FCAW and the results obtained by Goller and Paik for sampling periods of more than four hours may be unrealistically low.

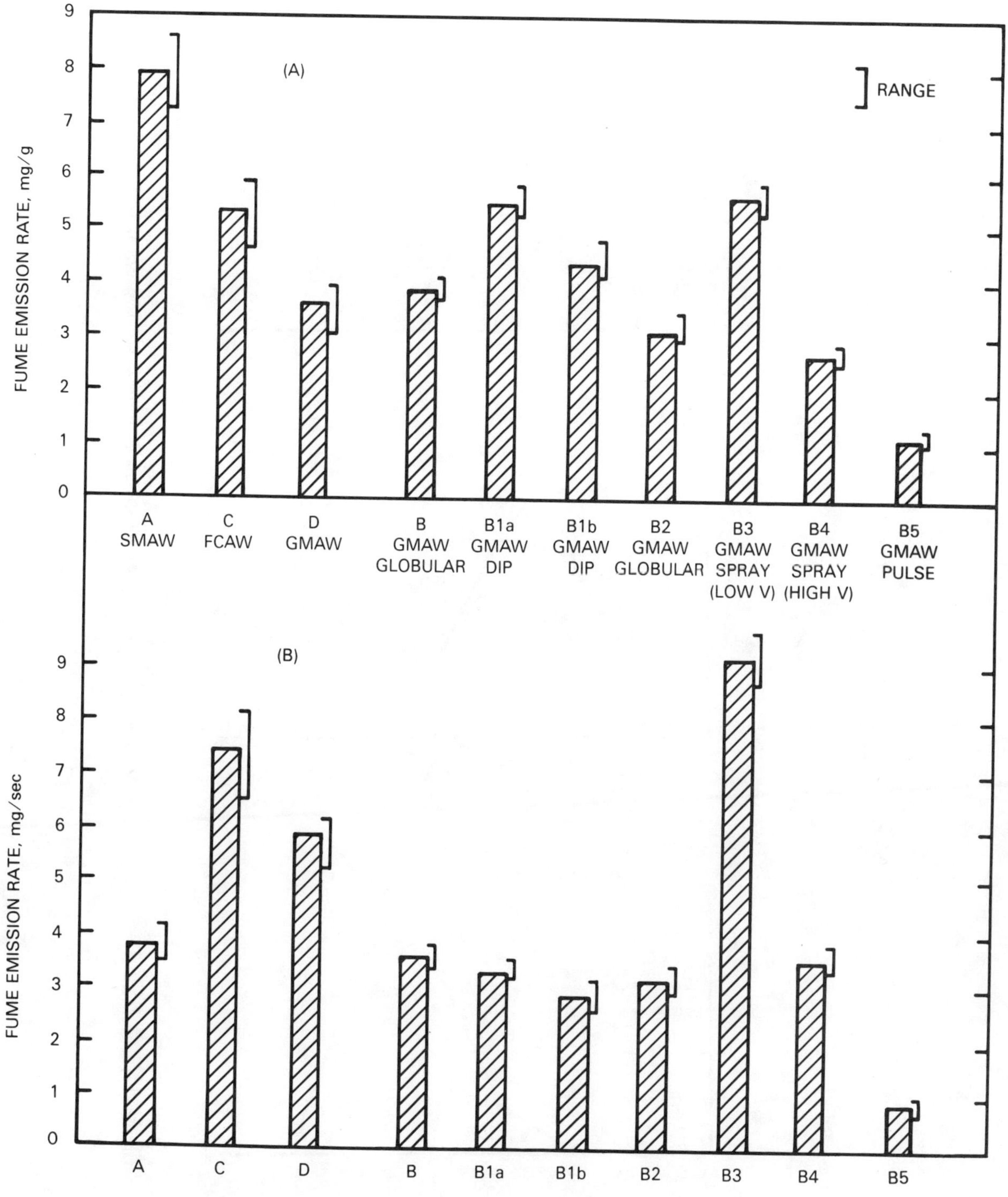

FUME EMISSION RATES: (A) EXPRESSED IN mg/g OF DEPOSITED METAL:
(B) EXPRESSED AS mg/sec OF ARCING

Redrawn from Moreton et al., (Ref. 99)

Figure 1 — Fume Emission Rates for Various Welding Processes

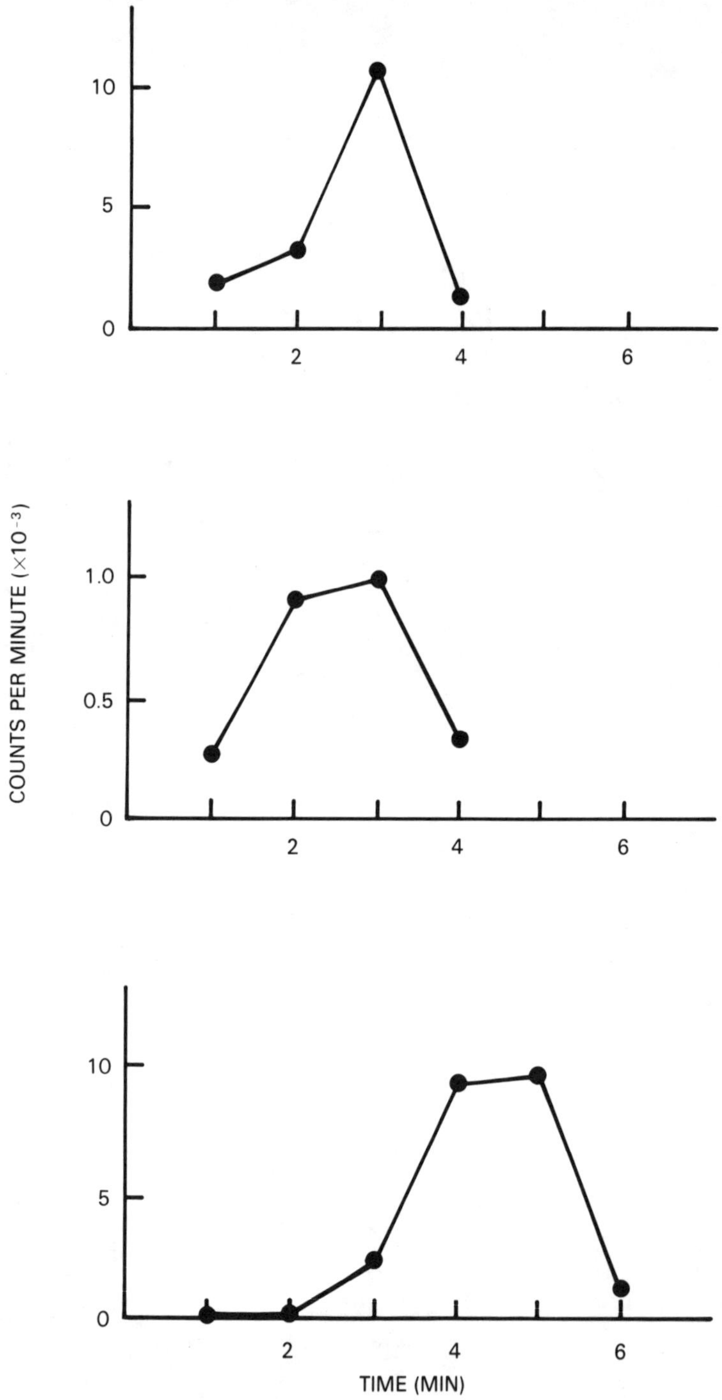

Redrawn from Glinsmann and Rosenthal. (Ref. 46)
Note: Measurements Made at One Minute Intervals. Results from Three Different Short-Term Welding Operations Are Shown.

Figure 2 — Fume Level Versus Time During SMAW of Steel

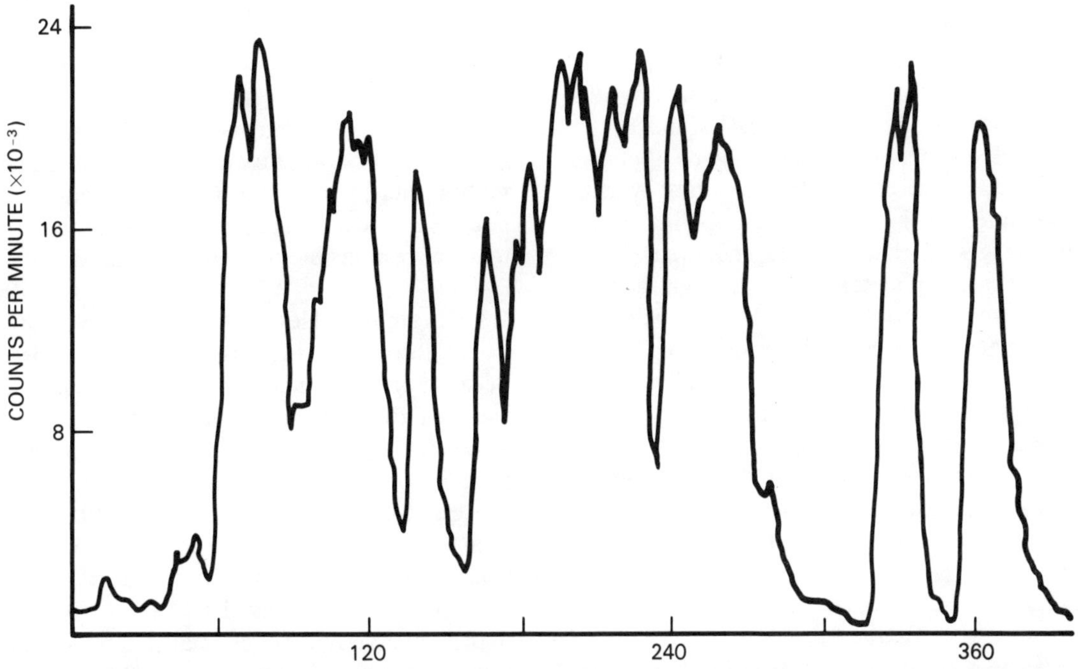

Redrawn from Glinsmann and Rosenthal. (Ref. 46)
Note: Measurements Made in One Second Intervals

Figure 3 — Buildup and Decay of Fume Levels During SMAW

1.2 Effects of Electrode Composition. The fume composition can vary with the FGR as well as with the formulation of the electrode coating. Tandon et al., (Ref. 151) found a strong correlation between the ratio of water-soluble hexavalent chromium to total chromium in the fume and the concentration of sodium and potassium in the flux. The iron content of the fumes was directly proportional, and the fluoride content was inversely proportional, to FGR's.

Limpel et al. (Ref. 81) found that the quantity of water-soluble fluorides, but not total fluorine, is considerably greater in fumes generated by electrodes containing potassium silicate than in those with sodium silicate.

You and Yan (Ref. 165) studied the mechanism by which hydrogen fluoride (HF) is generated during welding with basic electrodes. With a conventional electrode, 1 to 2 mg/m3 HF and traces of silicon tetrafluoride (SiF4) were found in the fumes. The quantities of HF and SiF4 increased with the concentration of silica (SiO2) in the electrode coating, but only HF increased with the moisture content. Analysis of fluorides released from fluxes with varying iron, silicon and fluoride contents, led to the conclusion that SiF4 is an intermediate in the generation of HF during welding with basic electrodes.

Manasova (Ref. 86) found that substitution of sodium silicate for potassium silicate in basic electrodes reduced fume generation by twenty five percent. The level of hexavalent chromium in the fume was directly dependent on the quantities of sodium silicate, potassium, and calcium levels in the electrode coating. On the basis of these findings as well as those of other investigators, an electrode was designed for welding of stainless steel of the type 18Cr8Ni2Mo which generated fifty three percent less total fume and twenty six percent less hexavalent chromium than did the conventional Czechoslovakian electrode E-B 427.

Olah and his co-workers (Refs. 116, 119, and 120) used X-ray fluorescence and neutron activation analyses to quantify a series of heavy metals (iron, manganese, chromium, nickel, vanadium, and molybdenum) in particles produced by SMAW of austentitic steel using a selection of coated electrodes. The relatively good agreement between results obtained with the two analytical techniques and the variation in heavy metal composition in fumes generated by different electrodes are shown in Tables 1A and B.

Thorne and Hewitt (Ref. 157) showed that the amount of metal released into fumes during brazing is directly related to the amount of bubbling in the flux. This suggested that bubbles encourage fume formation by carrying metal vapors through the flux barrier. The investigators postulated that bubbling resulted from heat-induced decomposition of fluoroborate fluxes and from hydrogen gas, which in the acidic environment of fluoroborate fluxes, could form by reaction of metals with the flux. This concept was evaluated by preheating the flux, which removed volatile gases and thereby reduced bubbling. Cadmium emissions were reduced threefold when preheated fluxes were used; however, a low concentration of fume was still present while brazing with preheated fluxes. These fumes were attributed, at least in part, to hydrogen gas which was found to evolve during brazing with preheated flux.

These experiments suggest that the evolution of fumes via bubbling may be reduced by raising the pH of the fluoroborate flux to reduce hydrogen gas formation and by lowering the concentrations of compounds (boric acid and potassium hydrogen fluoride) that are responsible for bubble formation.

Table 1A
Determination of Heavy Metals in Aerosols (%) Generated by SMAW Using X-Ray Fluorescence Analysis

Electrode	Fe	Mn	Cr	Ni	V	Ti	Mo
E-B 408	14.27	1.24	8.31	0.36		0.34	
E-B 415	6.23	15.20	1.54	0.10	—	—	—
E-B 427	7.78	6.81	4.42	1.47		0.01	0.11
E-B 428	6.17	6.46	5.26	1.59	0.88	2.15	0.15
E-B 445	7.90	9.15	2.02	2.88	—	—	
E-B 466	6.47	4.86	4.85	5.93		—	0.71
E-B 507	14.43	9.12	1.66		0.96	—	—
E-B 544	13.55	5.34		—	—	—	—

Olah and Tolgyessy, Ref. 120

Table 1B
Determination of Heavy Metals in Fumes Generated by SMAW (%) Activation Analysis with 14 MeV Neutrons

Electrode	Fe	Mn	Cr	V	Mo
E-B 121	24.70	3.45		—	—
E-B 415	6.01	16.17	1.29	—	—
E-B 427	7.85	7.34	4.11		0.10
E-B 428	6.80	7.05	4.92	0.82	0.13
E-B 445	8.04	10.25	1.98		
E-B 466	6.50	5.02	4.55	—	0.67
E-B 507	14.40	9.34	1.51	0.98	—
E-B 544	14.00	15.48		—	—

Olan and Tolgyessy, Ref. 120

1.3 Chromium. Hexavalent chromium, but not trivalent chromium, is a suspected human carcinogen. Because of this, much attention has focused on hexavalent chromium in welding fumes. The quantity of hexavalent chromium and the ratio of hexavalent to total chromium varies markedly with the welding process. Olah et al. (Refs. 117 and 119) showed that total chromium was three-to-fourfold higher, but hexavalent chromium was two orders of magnitude lower with argon-shielded GMAW of stainless steel than with SMAW. Zober et al. (Refs. 168 and 171) found that the amount of total and hexavalent chromium in fumes from stainless steel welding was much lower for GTAW and GMAW than for SMAW with rutile-coated or basic electrodes. Moreton et al. (Ref. 99) found that total chromium represented 5 percent of the fumes, whereas hexavalent chromium concentrations were 4.1 percent and 2.7 percent of the fumes, generated by SMAW and FCAW of stainless steel, respectively. The ratio of hexavalent chromium to total chromium was low in fumes generated by Ar-2 percent O2 shielded GMAW of stainless steel under a variety of welding conditions.

In 1983, Moreton et al. (Ref. 95) reported results from an interlaboratory round robin validation study of the Blakely and Zatka (Ref. 22) method for hexavalent chromium analysis (discussed in *Effects of Welding on Health, V*—Ref. 142). In response to this report, Dare et al. (Ref. 29) commented that the fuming perchloric acid method for dissolution of welding fumes may yield an underestimation of chromium due to loss of chromyl chloride. In addition, Dare et al. referred to the work of two groups of investigators (Refs. 53 and 156) which indicated that a short-lived form of hexavalent chromium appears in freshly formed GMAW fumes. These authors had shown that fumes collected by impingement in water had more hexavalent chromium than did those collected by filters. The hexavalent chromium concentrations peaked twenty seconds after fume formation and then began to decline. Within minutes, the hexavalent chromium levels were equivalent to those in fumes collected by filters.

Dare et al. remarked that this short-lived hexavalent chromium peak would not have been detected by sampling procedures used in the interlaboratory round robin study. Moreton et al. (Ref. 96) responded that the results obtained with fuming perchloric acid were in agreement with those of other techniques and that impingement sampling techniques would not be appropriate for on-site welding fume collection within the helmet.

Since that time, Zatka (Ref. 166) published a modified method for determination of hexavalent chromium which resulted from reports that small but significant concentrations of hexavalent chromium form during the digestion of trivalent chromium in the hot alkaline solutions used in the analytical method. Zatka (Ref. 166) demonstrated that the formation of hexavalent chromium during alkaline digestion can be suppressed by the addition of magnesium salts to the alkaline digestion medium.

In regard to the observations of Thomsen and Stern (Ref. 156) and of Gray et al. (Ref. 53) of a short-lived hexavalent chromium species in freshly formed welding fumes, Zatka (Ref. 166) suggested that this fume aging phenomenon may result from a chemical reaction within the impingement collector fluid rather than by a reaction within the fume solids. Suziki and Serita (Ref. 149) developed a method for determination of water-soluble trivalent and hexavalent chromium by anion exchange high-pressure liquid chromatography. Trivalent chromium was chelated with ethylene diamine tetraacetic acid (EDTA) before analysis by liquid chromatography. The recovery from columns of solutions containing pure compounds [chromic chloride ($CrCl_3 \cdot 6H_2O$) and potassium dichromate ($K_2Cr_2O_7$)] or welding fumes was close to 100 percent and detection by UV spectrophotometry and atomic absorption were equally effective.

1.4 Barium. Because soluble barium salts are toxic, whereas insoluble salts are relatively inert, Dare et al. (Ref. 28) examined total and soluble barium levels in fumes released from three types of electrodes with barium fluxes. Total barium was determined in perchloric acid extracts and soluble barium was determined in distilled water or 0.1 M hydrochloric acid (HCl) extracts of fumes. The results are shown in Table 2. The barium in fumes from all electrodes was substantially soluble in both dilute acid and water, although the rate at which barium compounds were extracted from the fumes differed. The investigators stressed that measures should be taken to control exposures to fumes from electrodes containing barium because of their high content of soluble barium.

In response to the article of Dare et al. (Ref. 28), Moreton and Jenkins (Ref. 98) published data showing that fume concentrations of soluble and total barium differ substantially for various flux-cored self-shielded electrodes. The amount of total barium ranged from 0.1 to 34 percent and the proportion of soluble to total barium also varied widely. According to Moreton and Jenkins, the concentration of barium in the fume varies with wire diameter and process conditions (e.g. voltage, wire-feed speed). Since neither the quantity of barium present in the electrodes nor the conditions of fume generation were reported by Moreton and Jenkins (Table 3), their data cannot be compared directly with those of Dare et al. (Ref. 28).

1.5 Particles. Welding fumes are primarily composed of microscopic particles which, depending on their size and shape, may become deposited in different sections of the respiratory tract. Spherical particles larger than 5 μm in aerodynamic diameter are usually removed from the airstream in the upper respiratory tract and are expelled from the lungs. Particles between 0.1 and 5 μm in diameter are considered to be respirable; they can be inhaled and retained within the lower respiratory tract. Particles smaller than 0.1 μm are generally not removed from the airstream during the respiratory cycle, and they exit the lung with exhaled air.

Welding fume particles tend to be of respirable size. They are generally spherical, although the regularity of the surface may vary with the welding process and electrode. Particles may be present in fumes as single entities or as agglomerates or chains of varying lengths (Refs. 107, 118 and 126).

Mean particle sizes vary with the welding or metal cutting process. Olah et al. found that particle sizes in fumes produced by welding austentitic steel ranged from 3 to 5 μm for SMAW with alkaline electrodes, from 0.8 to 0.9 μm for SMAW with rutile-coated electrodes, and from 0.1 to 0.2 μm for argon-shielded GMAW with austentitic welding wires (Ref. 118). Particles released during plasma arc cutting of steel were primarily composed of magnetite (Fe_3O_4) and were 0.8 to 0.9 μm in diameter. The particles generated by oxygen cutting of mild steel had a similar composition, but were much smaller in size (0.4 μm). Significant concentrations of chromium and nickel were present in particles released during plasma arc cutting of stainless steel. Particles generated by cut-

Table 2
Content and Solubility of Barium in Welding Fume

	Electrode		
	Innershield	Soudofonte B12	Soudofonte B1
% Barium in flux	-40	-40	-20
Barium salt in flux	Barium fluoride	Barium carbonate	Barium carbonate
% Barium in fume	16	33	16
Proportion extracted in 0.1 M HCl	90% in 60 min	100% in 5 min	100% in 5 min
Proporation extracted in water	70% in 120 min	90% in 180 min	100% in 120 min

Data from Dare et al., Ref. 28

ting of aluminum and copper were approximately 0.5 μm and 0.14 μm in diameter and contained primarily aluminum oxide ($Al2O3$) and copper oxide ($Cu2O$), respectively (Ref. 108).

Deposition of particles in the lungs is dependent not only on their size and shape, but is also influenced by their electrostatic charge. Johnston et al. (Ref. 68) measured the electric charge on aerosols in a variety of workplaces and found that the median electrostatic charge of 0.2 μm particles in fumes produced by GMAW and SMAW was relatively low. This was thought to be caused by "charge-aging" whereby free ions in the air progressively neutralize airborne dust.

With scanning electron microscopy, welding particles appear to be granular and to have a dense inner core surrounded by a transparent outer shell. The appearance of the outer shell differs in particles produced by SMAW and by submerged arc welding (Ref. 107). The regularity of the particle surface, core size, and texture varied with the electrode and welding conditions (Refs. 107 and 126).

The outer surface of particles generated by SMAW were found to contain iron oxide, silicon dioxide, silicates, and titanium. The particle cores, as revealed by ion sputtering, were composed primarily of manganese oxide (Ref. 107).

Golovatyuk et al. (Ref. 48) examined particles generated by CO2-shielded welding of low alloy steel with a coated electrode and a flux cored electrode. Fe, Ca, Ti, F, Si, and traces of Cu, Ba, and S were detected by mass spectroscopy of solid constituents in fumes produced by both types of electrodes. X-ray diffraction analysis revealed magnetite (Fe3O4), chromic oxide (Cr2O3), and ferrous oxide (FeO) in particles from the flux cored electrode and Fe3O4, manganous oxide (MnO), and SiO2 in those from the coated electrode.

X-ray diffraction analysis of fumes generated by SMAW and argon-shielded GMAW of austentitic steel revealed ferric oxide (Fe2O3), Fe3O4, aluminum chromium iron oxide [(FeCrAl)2O4], and trace quantities of Al2O3 (Ref. 118).

Using X-ray powder diffraction, Tanninen (Ref. 154) analyzed fumes produced by SMAW of mild steel. The sample contained Fe3O4. Although the certainty of the detection method was limited, hematite (Fe2O3), manganese iron oxide, and calcium or iron silicates also appeared to be present. Potassium calcium fluoride (KCaF3) was found, and there were indications that potassium iron fluoride (K3FeF6) was also present. Tanninen (Ref. 155) also used X-ray spectrometry to examine the valence of iron in welding fumes generated by SMAW and GMAW of mild and stainless steel. Only fumes generated by SMAW of stainless steel contained significant amounts of Fe3+.

As part of a multi-disciplinary project to investigate chemical, physical and toxicological properties of fumes from five electrodes, Tandon et al. (Ref. 151) determined concentrations of nineteen elements in the flux and fume from three types of hardfacing (medium chromium, high chromium, and high mag-

Table 3
Analysis of Amounts of Barium in Fume

Self-Shielded Innershield Wires (Code)	% Barium in Fume: Water Soluble Ba (37°C, 1h)	% Barium in Fume: Total Ba (X-ray fluorescence)
A	20.3	22.0
B	13.9	33.8
C	–	30.6
D	–	0.1
E	–	5.9

Moreton and Jenkins, Ref. 97

nesium contents) and two types of high-strength, low-alloy steel flux-coated electrodes. The only crystalline compounds detected in the fumes by X-ray diffraction were Fe_3O_4, potassium chromate (K_2CrO_4), calcium fluoride (CaF_2), and sodium fluoride (NaF). Crystalline silica or metal silicates were not detected in any of the fumes studied. Using a cascade impactor to fractionate fumes from a hardfacing electrode with a high magnesium content, Tandon et al. (Ref. 152) found that ninety two to ninety six of the fume particles were less than 7 μm in diameter. Nine elements (F, Na, Mg, K, Ca, Cr, Mn, Fe, and Ni) were determined in each fraction and found to be equally distributed among the particles in all size ranges.

Using X-ray photoelectron spectroscopy and other techniques, Tandon et al. (Ref. 153) measured the concentrations of fourteen elements on the surface of particles in fumes generated by SMAW of stainless steel. NaF and potassium fluoride (KF) comprised fifty percent of the particle surface; SiO_2 and soluble chromate accounted for an additional thirty percent and eight percent of the surface, respectively. Thermal methods of analysis detected Fe_3O_4, K_2CrO_4, and possibly sodium chromate (Na_2CrO_4). According to the authors, the remainder of the particle surface was probably occupied by transition metal oxides, hydroxides, or silicates. The surface remaining after soluble fluorides and chromates were removed by washing, contained primarily SiO_2 (about sixty percent) and some transition-metal oxides, silicates and fluorides.

The manganese concentration was as high as forty percent in fumes generated by hardfacing with basic electrodes containing large quantities of manganese. The MnO content of these fumes increased with increasing concentrations of carbon dioxide (CO_2) or by the addition of oxygen to the argon shield gas. X-ray diffraction analysis revealed that manganese was present as iron manganese oxide ($MnFe_2O_4$) and potassium manganate III ($KMnO_2$) in the fumes liberated from coated electrodes (Ref. 65).

2. Gases

Ozone, carbon monoxide, carbon dioxide and nitrogen oxides are the principal gases generated by welding. Other vapors and gases may evolve when surfaces contaminated with paints, greases or other surface coatings are welded or when degreasing solvents are present on the weld metal or as vapors in the vicinity of the welding operation. These gases and vapors are discussed in more detail in preceeding sections (see "Production Coatings" and "Degreasing Agents").

Carbon monoxide and carbon dioxide arise from combustion of carbonaceous materials and are present during oxyfuel gas welding and CO_2-shielded GMAW. They may be generated by combustion of carbonates or organic materials in the electrode coating or on the metal surface. Carbon dioxide is a simple asphyxiant, while carbon monoxide interferes with oxygen transport by the blood.

Nitrogen oxides (NOX) arise from thermal oxidation of molecular nitrogen in air. NOX may be generated during most welding processes but the concentrations of NOX are greatest during oxyfuel gas welding and plasma arc cutting. Nitrogen oxides can irritate the eyes, nose, and throat and in very high concentrations, can cause pulmonary edema and death. Chronic exposure to NOX may cause reduced pulmonary function.

Ozone is generated by exposure of oxygen to ultraviolet light in the wavelength range 185 and 220 nm. It is a severe respiratory irritant; exposure to levels above 0.3 ppm can cause extreme discomfort while exposure to 10 ppm for several hours can cause pulmonary edema. Ozone is unstable in air and its decomposition is accelerated by metal oxide fumes. Therefore, significant quantities of ozone are generally not associated with welding processes, such as SMAW and FCAW, which generate large quantities of fumes.

Nemcova (Refs. 108 and 109) used a portable sampling chamber to measure the concentrations of carbon dioxide, carbon monoxide, ozone, nitrogen oxides, and hydrogen fluoride generated by different welding and cutting procedures. Hydrogen fluoride and nitrogen oxide levels generated by SMAW with basic electrodes and submerged arc welding with several different electrodes were well below Czechoslovakian and U.S. TLV's. Relatively high concentrations of nitrogen oxides were generated during plasma arc cutting of steel, aluminum and copper, and argon-shielded GTAW of copper and aluminum. Of the welding processes examined, ozone levels were highest during plasma arc cutting (Table 4).

Using an open chamber designed to measure welding gases, Tigges (Ref. 158) determined that the quantities of nitrogen oxides, carbon monoxide, carbon dioxide, and methane released during SMAW with thirty three different electrodes varied considerably. For example, the volume of nitrogen oxides ranged from 9 to 100 ml/min with different electrodes. The total quantities of individual gases varied

Table 4
Levels of NO_x and Ozone Generated During Welding and Cutting

Process	Metal	NO_x (mg/m³)	Ozone (mg/m³)
Argon-shielded GTAW	copper	3.3	NG
Argon-shielded GTAW	aluminum	3.8	NG
SMAW (basic electrode)	NG	0.1	0.04
Submerged arc welding	NG	<0.005	0.01
Plasma arc cutting	steel/SS	7.8	0.08
Plasma arc cutting	steel/MS	7.5	0.08
Plasma arc cutting	aluminum	5.8	0.08
Plasma arc cutting	copper	7.3	0.08
Oxyfuel cutting	steel/MS	1.4	0.04
Resistance Seam Welding (copper electrode)		0.5	0.006

NG—data not given; SS—stainless steel; MS—mild steel
Data from Nemcova, Ref. 108 and Ref. 109

substantially even for electrodes with similar chemical compositions.

Fog and Ritz (Ref. 43) developed a piezoelectric detector for monitoring ozone in the workplace. The detector utilized a crystal coated with 1,4-polybutadiene. Ozone concentrations were measured by changes in the frequency of the crystal which result from ozone-induced alteration of the polybutadiene coat. The lower limit of detection was less than 10 ppb ozone. The ozone produced by GMAW and GTAW of aluminum measured outside and within the welders helmet, as well as in various positions with respect to the arc, were measured simultaneously with the piezoelectric detector, dosimeter and gas detector. The results of the three detectors were in relatively good agreement.

3. Electromagnetic Radiation

Electromagnetic radiation in the ultraviolet (UV), visible, and infrared (IR) regions of the spectrum is produced during welding. UV radiation can cause "welder's flash" and skin burn. Intense visible light of certain wavelengths can cause retinal injury and IR may cause cataracts and thermal damage to the cornea and other tissues.

Using a rapid scan spectrometer Eriksen (Ref. 37) measured optical radiation from the UV through visible wavelengths (200 to 800 nm) emitted during GMAW of aluminum with an aluminum/magnesium electrode. Spectra were obtained in five millisecond intervals before and after ignition of the arc. Eriksen demonstrated that a burst of high level UV radiation occurs during the initial phase of arc ignition and lasts less than fifty milliseconds after the arc is struck. The intensity of this UV overshoot was more than ten times that of the UV light emitted when the arc was burning continuously. Calculations performed according to ACGIH guidelines indicated that with a welding current of 300 A, the unprotected eye at a distance of 0.5 meters may suffer "welder's flash" after exposure to radiation from only one ignition of the welding torch.

Okuno (Ref. 115) compared radiation spectra in the wavelength region 200–1000 nm produced by a variety of welding conditions. He examined spectra produced by SMAW of mild steel with titanium-coated (ilmenite) or low hydrogen electrodes and with argon- or CO2-shielded GMAW and GTAW of stainless steel, aluminum, and mild steel.

Reproducible and characteristic spectra were produced by each set of welding conditions although there were substantial fluctuations in intensity. For any one set of welding conditions, the intensity increased with the wire diameter, arc current, and voltage.

The most prominent spectral lines generated by GTAW of stainless steel were in the near IR (700–1000 nm) region. The spectrum produced by argon-shielded GMAW of aluminum was characteristic of emissions generated by the base metal and the argon shield gas and, unlike other welding conditions, the UV spectrum was much more intense than the visible spectrum. Most of the prominent lines observed in the UV region were emissions from magnesium which accounts for a large part of the UV radiation from welding aluminum when magnesium is present in the electrode or the base metal. This is in accord with previous reports (Ref. 16).

3.1 Protective Eye Wear. Heat can be passed to the face of the welder by direct transmission of infrared radiation through the goggles. In addition, IR can be absorbed by the goggles themselves and then passed to the face of the welder by secondary transmission of absorbed heat. Reflective materials on the outer surface of the goggles can reduce heat transfer by reflection of IR radiation. The direct and secondary transmission of IR radiation through several types of goggles and the ability of metallicized surfaces to reduce this transmission was examined by Felzmann (Ref. 42).

Eighteen types of rectangular goggles (ten with aluminum reflective surfaces, one with a copper surface), five types of round goggles (two with aluminum surfaces), and four goggles that were tinted blue (two with aluminum and two with copper surfaces) were tested. No remarkable differences were seen in the transmission of IR radiation among goggles within each category. The blue tint did not provide any protection against heat transmission.

Since sufficient heat must be absorbed by the goggles before secondary heat transmission occurs, direct and secondary IR transmission can be distinguished by measuring heat transmission with time. Felzmann found that both aluminum and copper coats effectively reduced IR transmission, but the aluminum surface was less efficient at reducing secondary transmission of absorbed heat. Felzmann speculated that this was due to heat absorption by an outer coating composed of an unknown substance which was applied by the manufacturer to protect the aluminum coating from physical damage.

Mosely (Ref. 103) examined transmission of ultraviolet and visible radiation between 200 to 800 nm through two types of welding goggles. When fitted with a filter rated for use with flux in GMAW of aluminum and aluminum-magnesium alloys, lead welding, and oxyacetylene cutting, one type of goggle completely attenuated UV radiation. Another filter, intended for use while welding with the processes listed above without flux, fitted to the other type of goggle transmitted a small amount of long wave ultraviolet radiation. The front filters of both goggles, designed to protect the filter from splatter and enhance mechanical protection, had a negligible effect on the transmission of light above 200 nm. Limited UV protection was provided by the underlying British Standard 679 welding filter. The clear backing lens on the second goggle with the lift-up front provided little protection from UV radiation.

3.2 Heat. The pain threshold for heat on the skin is 0.2 to 0.3 watts/cm2. Radiation of this intensity can cause intense pain or even skin burns within one to four minutes. Irradiation with 0.1 to 0.2 watts/cm2 causes the skin to feel hot, while 0.005 to 0.01 watts/cm2 is the lower threshold of heat sensitivity and does not cause discomfort. Richter and Ruess (Ref. 134) studied the exposure to radiant heat of welders using GTAW with and without preheating and grey cast iron welding with preheating. The amount of infrared radiation that would strike different parts of the body of welders working in different positions was measured. Their results indicated that special heat protection was not necessary for GTAW without preheating. However, the pain threshold of 0.2 watts/cm2 was surpassed in several instances during grey cast iron welding and GTAW with preheating of the metal.

Artemev (Ref. 10) tested physiological effects of various fabrics and materials considered for use in clothing designed to protect welders against burns from sparks and spatter from molten metal. The sleeves, trousers and the front part of the jackets constructed for these tests were made of fireproof fabrics. Body temperature, skin pH and temperature, heart rate, and skin microorganisms were measured before and after performing work.

Clothing composed of artificial leather, linen or a mixture of linen and other fibers did not cause substantial changes in any of the parameters tested and were recommended by Artemev for wide use at temperatures between 21 and 33°C. A synthetic material denoted as phenylone caused changes in the pH balance of the skin. Vinyl leather caused considerable changes in the body temperature, heart rate, skin microorganisms, and skin pH and hence was only recommended for use when absolutely necessary and only under optimal conditions of temperature and humidity in the work environment.

4. Production Coatings

Welding of metals coated with paints or primers can introduce hazards distinct from those inherent in welding of clean metal. Doorgeest (Refs. 32 and 33) cautioned that highly toxic gases and vapors can be released during welding of metals coated with isocyanate-containing paints or anti-fouling paints. The latter may contain metals such as tin, mercury and copper. He recommended that, prior to welding, the paint layer be removed to a width at least 15 cm or more on all sides of the area to be heated. Similarly, McMillan (Ref. 91) cautioned that oils should be removed from metals before they are welded. Degreasing agents or their vapors should not be in the vicinity of the weld because of the potential for photochemical conversion to highly toxic compounds.

Moreton (Ref. 97) and McMillan (Ref. 91) reviewed the hazards of welding coated or contaminated surfaces. Depending on the surface coating, a range of toxic metals and organic materials may evolve when heat is applied. Welding of metals coated with greases or paints containing synthetic chemical polymeric binders may emit highly toxic chemicals such as carbon monoxide, benzene, phosgene, hydrogen cyanide, polycyclic aromatic hydrocarbons and nitrogen dioxide. Welding of metals coated with zinc primers may release sufficient zinc to cause metal fume fever, and welding of metal coated with cadmium may cause serious acute effects and even death.

Standard procedures for assessing the toxicity of fumes released from flame cutting or welding of primed metals were recommended by representatives of the welding and paint industries in 1964 and revised in 1968 (Ref. 8). In the tests, metal pieces of a specified size, coated with a primer applied according to the manufacturer's instructions, is welded or cut. Thermal degradation products are captured, analyzed, and their concentrations compared with Threshold Limit Values.

According to Moreton (Ref. 97), the recommended procedures are outmoded and insufficient for realistic appraisal of the hazards associated with welding and cutting of primed materials. Because the tests are inadequately described, results cannot be duplicated among laboratories. According to the specifications, the samples need not be collected in the area of the plume to which welders may be exposed. The total quantities of fumes and gases released during welding are not considered. The size of the weld and duration of sampling time are unnecessarily large and impart extra costs to the method. The selection of degradation products to be assessed is at the discretion of the analytical laboratory, and thus, important chemicals may not be considered. Finally, the analytical procedures are not standardized. Moreton stressed the urgent need to adequately define procedures for these tests. Collaborative interlaboratory tests are necessary to authenticate test methods.

In summary, improvements in these procedures should be made which introduce reproducibility between testing laboratories, require less material, use shorter sampling times, and more specifically define test methods. New standards should be developed which offer guidance on the types and concentrations of pollutants likely to be present. The welding and cutting specifications should be based on the requirements of the analytical techniques and should be designed to provide sufficient sample for analysis.

5. Degreasing Agents

Fumes from degreasing agents or paints can present a major hazard in welding shops (Refs. 9 and 91). Chlorinated hydrocarbon solvents such as trichloroethylene, perchloroethylene and 1,1,1-trichloroethane can decompose in the presence of ultraviolet light into highly toxic compounds such as phosgene and dichloroacetyl chloride. Phosgene is a severe lung irritant capable of causing discomfort at 0.3 ppm and pulmonary edema at exposure levels of 10 ppm.

Knudsen and Bjerre (Ref. 74) developed a mathematical model for approximating the health risk of photochemical oxidation products during electric arc welding at a distance of 30 cm from the arc. Of the three halocarbons examined by this model, trichlorofluoromethane (CCl_3F) was found to be far more dangerous than carbon tetrachloride (CCl_4) or dichlorodifluoromethane (CCl_2F_2). However, even the latter two present a serious health risk when welding with systems that are significant sources of UV radiation, such as argon-shielded GMAW of aluminum.

Photochemical decomposition products resulting from reaction of degreasing agents with UV light produced by GMAW may have been responsible for an incident described by Ross (Ref. 135) in which eleven welders complained of coughing, breathlessness, chest tightness, and irritation of the throat and eyes. The source of the irritation was unknown. Both GMAW and SMAW of stainless steel were being used. The metal pieces were degreased before welding by placement in tanks containing either trichloroethylene or 1,1,1-trichloroethane. The degreasing operation was on the shop floor in the vicinity of the welders where there was good general ventilation.

Ambient air tests indicated that concentrations of trichloroethylene and trichloroethane in the work environment were well below their respective TLV's and were too low to have caused noticeable health effects. Records maintained by the welders indicated that the irritant was present during GMAW, but not when SMAW was used. This was confirmed by lung function tests before and after the work shift which showed that the forced expiratory volume (FEV) was reduced after GMAW only. Since substantial quantities of ultraviolet radiation are produced during GMAW only, photochemical conversion of chemical vapors to highly irritating substances such as phosgene and dichloroacetyl chloride was suspected. Phosgene could not be detected in the air and a method for measuring dichloroacetyl chloride was not available; thus, this hypothesis could not be substantiated. However, as a result of the presumed association, the welding operation was moved to an area distant from sources of chlorinated hydrocarbons, which successfully resolved the problem.

A Swedish welder developed pulmonary edema after performing argon/20 percent CO_2-shielded

GMAW of mild steel in an environment containing high levels of 1,1,1-trichloroethane used as a degreasing agent. Hallne (Ref. 56) reconstructed the conditions to which the affected welder was exposed to investigate whether toxic gases were generated in the work environment. Levels of phosgene, ozone, NOX, CO, CO2 and HCl were determined during GMAW in an environment deliberately contaminated with trichloroethane. Airborne phosgene concentrations as high as 0.4 ppm (TLV = 0.1 ppm) and hydrogen chloride concentrations as high as 37 ppm (TLV = 5 ppm) were measured while welding in a glove box where the trichloroethane level was 1000 ppm. Only negligible quantities of these gases were produced by welding in the absence of trichloroethane.

Similar tests were performed using the same welding conditions in an open workshop with good ventilation. None of the gases tested were present in concentrations sufficient to have caused the poisoning. Hallne concluded that the results did not provide any explanation for the poisoning assumed to be related to welding exposures.

6. Noise

Impulse noise may be more harmful to hearing than continuous steady state noise of the same energy level (Ref. 86). Lahti et al. (Ref. 78) defined the impulse level of noise as the difference between the peak value and the root mean square of the slow time-weighted value. Using this definition, noise is considered as impulse noise if the difference between peak levels and the slow time-weighted value is greater than 15 decibels. Tests performed at a shipyard plate welding workshop indicated that grinding and carbon arc work do not produce impulse noise (as defined above) while welding, and especially GMAW, generated high impulse noise levels.

Section Two Effects of Welding on Human Health

Adverse health effects may be associated with physical and chemical agents generated by welding, including gases, fumes, radiation, and noise. While a great deal of research has been performed on the health effects of welding, much is unknown about how these various agents affect the body. Understanding health effects is essential to the design of protective clothing and equipment and to the development of welding processes that emit minimally hazardous agents. Described in this section are health reports which appeared in the published literature from July 1984 through December 1985.

7. Respiratory Tract

The respiratory tract is the primary route by which welding fumes and gases enter the body. When inhaled, particles and irritant gases such as NO, NO2, HCl, ozone, and phosgene can elicit an inflammatory response. The intensity of the response is dependent on the dose (quantity of inhaled material), physical/chemical properties of the irritant, and the overall health status of the exposed person. A diverse population of cells migrate into the lung in response to irritatation. These cells proliferate mediators, causing inflammation of lung tissue which may result in cough, chest tightness, and pain. Severe respiratory irritants, such as ozone, phosgene, or cadmium, can cause accumulation of fluids in the lung with chemical pneumonitis or pulmonary edema developing hours after exposure. Chronic exposure to low concentrations of irritants may result in emphysema and bronchitis, damage to the air sacs (alveoli) and airways (bronchi and bronchioli), respectively. The association between exposure to welding fumes and the developmemt of chronic lung diseases such as bronchitis, emphysema, or lung cancer has not been clearly established.

Much of the particulate material deposited in the respiratory tract is engulfed by alveolar macrophages (scavenger cells). Both particle-laden macrophages and free dust particles are removed from the respiratory tract by coughing and the natural movement of surface fluids lining the air passages. Some of the dust and particle-laden macrophages are deposited in pockets in the lungs where they can remain for extended periods of time.

Extensive accumulation of dust particles in the alveoli can prevent expansion of the air spaces during inhalation and may thus interfere with lung function. In addition, some dusts induce the multiplication of fibroblasts or connective tissue cells along the interior surfaces of the lung which can result in deposits of collagen fibers or scar tissue. The pockets of particles and collagen fibers produce the small, round or irregular shadows visible in X-rays of "arc welders" lungs. By distorting lung structure and obliterating air spaces, the scar tissue can interfere with lung function. Other particulates, such as iron oxide, are relatively inert and have only minor effects on the lung. In their presence, fiber formation may be negligible or minimal (Ref. 147), and the anatomical integrity of air spaces remains undisturbed. The lesions caused by these nonfibrogenic particles are frequently reversible.

Pneumoconiosis is a general term for the "the accumulation of dust in the lungs and the tissue reaction to its presence" (Ref. 64). It is usually detected as opacities in chest X-rays, but can also be diagnosed by microscopic examination of lung biopsy material. The latter technique enables determination of the characteristics of the particle deposits which is useful for determining the prognosis of the lung condition.

To facilitate the exchange of information between lung specialists, an international classification system for pneumoconiosis was developed under the aegis of the International Labour Office (Ref. 64). According to their definition, a "non-collagenous" pneumoconiosis results from the accumulation of inert dusts such as ferric oxide. Characteristics of this condition are (1) the anatomic integrity of the lung architecture remains intact, (2) there is no scar tissue, and (3) it is potentially reversible. Several studies have characterized the pneumoconiosis, sometimes observed in arc welders, as described above (Refs. 130, 131, 137, 142, 144 and 147). However, there is substantial variation in the fibrogenicity of dusts emanating from different welding processes, and it is thus possible that arc welders' pneumoconiosis may not always be benign.

Spacilova and Hykes (Ref. 142) assessed pulmonary changes in 85 Czechoslovakian arc welders who had been admitted to their clinic with abnormal chest X-rays. Examination of lung biopsy material from five of the welders revealed pockets of dust containing primarily iron. Some fibrous tissue was present, but the structural integrity of the alveoli was undisturbed. With the exception of minor changes, lung function tests were within normal limits.

Follow-up examinations administered to forty four of the welders an average of 6.6 years later revealed no substantial changes with time. None of the welders developed indications of massive progressive fibrosis. In four welders who had excessive exposure to welding fumes, the radio-opaque shadows regressed after they permanently stopped welding. Spacilova and Hykes concluded that the X-ray findings in arc welders were not indicative of a pathological condition.

Morgenroth and Verhagen-Schroeter (Ref. 101) examined biopsies of lung tissue taken from seven welders with pneumoconiosis. Both needle shaped and round particles, 0.05 to 0.25 mm in diameter, were found engulfed by macrophages. Energy dispersive X-ray analyses indicated that the needle shaped particles were composed of aluminum silicates and the round particles contained primarily iron. Substantial quantities of phosphorus and lesser quantities of other elements were also found in particles in the lungs. The close proximity of macrophages with engulfed particles and fibroblasts supported the theory that macrophages proliferate mediators which stimulate fibroblast division and fibrosis.

Fibrosis, emphysema, and other respiratory diseases may cause measurable changes in lung function. Pulmonary function tests are used to detect disease processes which restrict lung expansion or reduce pulmonary elasticity, or both. These tests measure the air volume which can be inhaled or expelled either forcefully or under normal breathing conditions. The types of measurements made in lung function tests include the forced vital capacity (FVC), the forced expiratory volume during the first second of exhalation (FEV_1), and the maximal expiratory flow rate (MEF).

The FVC is a measure of the volume of air that can be expelled forcefully following a maximal inspiration. This measure reflects resistance to airflow in the lungs. The FEV_1 and MEF measure the percent of the total volume or the rate of air expelled during a portion of the respiratory cycle. These measurements all serve as indicators of resistance to air flow. [The reader is referred to Appendix C of the *Effects of Welding on Health,* Volume I (Ref. 161) for a detailed explanation of the terms used in this section.]

Lung function tests are frequently used to monitor industrial workers for damage to the respiratory tract, such as changes in pulmonary compliance and airway obstruction. These measurements are not always sensitive to early changes in the lungs and irreversible damage may occur before noticeable reduction in pulmonary function is detected.

Sjogren and Ulfavorson (Ref. 139) examined lung function and the incidence of respiratory symptoms in 64 gas metal arc and gas tungsten arc welders of aluminum, 46 shielded metal arc welders of stainless steel, and 149 welders of railroad tracks who performed SMAW with basic electrodes as well as with electrodes containing 3.2 percent chromium. Breathing zone concentrations of ozone were frequently above the Swedish standard of 0.1 ppm during GMAW of aluminum, and chromium levels were above the standard of 20 ug/m3 during SMAW of stainless steel. Concentrations of nitrogen oxides were usually below the standard of 5 ppm with all welding methods.

There were no differences in FVC and FEV_1 between welders and controls. Welders complained of greater occurrences of cough, phlegm, and respiratory irritation than did nonwelders. Among welders performing GMAW of aluminum, the frequency of respiratory symptoms increased with increasing exposure to ozone. There was no relationship between exposure to total particulates and respiratory symptoms, but a nonsignificant tendency for the incidence of respiratory symptoms to increase with exposure to chromium was observed. The investigators concluded that ozone, rather than aluminum particulates, was responsible for respiratory symptoms in gas metal arc welders of aluminum.

For stainless steel and railroad track welders, respiratory symptoms correlated better with exposure to chromium than to total particulates. The investigators suggested that variations in plant ventilation may have accounted for differences between their findings that welding emissions did not cause changes in pulmonary function and other studies which showed the opposite effect.

Mur et al. (Refs. 105 and 106) determined the incidence of abnormalties in pulmonary function and symptoms of respiratory disease in 536 arc welders and 427 control workers employed at three French factories. SMAW of mild steel was the most common welding method used at one of the plants while GMAW of mild steel and aluminum were used almost exclusively at the remaining plants.

Industrial hygiene measurements of welding fume components and gases were made at thirty representative welding sites within these factories (Ref. 30). TLV's were exceeded for at least one substance in twenty seven of the thirty work areas. In most cases, source ventilation was not used. However, in the three work posts where local exhaust was available, one or more of the measurements were still above the TLV. Ozone was exceeded in one work area during argon-shielded GMAW of aluminum, and carbon monoxide was elevated in one area where carbon dioxide shielded GMAW was performed. Pulmonary function and X-ray abnormalities did not correlate with the metals welded or the welding procedures.

Recurrent bronchitis and breathlessness were more frequent among shielded metal arc welders than gas metal arc welders. Shielded metal arc welders also had significant reductions in the ventilatory functions MEF and FVC. A significant reduction of lung diffusion capacity for CO and in the lung diffusion coefficient was observed among gas metal arc welders at one factory.

At this same plant, welders who worked in confined spaces had lower maximal expiratory flow at twenty five percent vital capacity than did welders who worked in well ventilated areas. Radiological abnormalities were more frequent in welders who worked in confined spaces, and also in shielded metal arc welders. As expected, tobacco smoking had a marked adverse effect on respiratory symptoms and lung function in all workers (welders and nonwelding controls). Smoking apparently interacted with welding emissions since CO lung transfer was more impaired in smoking welders than in smoking controls.

In summary, the effects of welding exposures on lung function and chest X-rays differed among the plants studied, and greater changes were observed in the lung diffusion capacities for CO than in spirometric measures of lung volumes. Because radiographic and lung function abnormalities were observed among the welders in this epidemiologic study, Mur et al. (Ref. 105) concluded that industrial hygiene measures at welding workshops should be enforced and chest X-rays and CO transfer tests should be used to monitor the respiratory state of welders.

Schneider (Ref. 137) compared the pulmonary status of 433 welders and 421 nonwelders who worked at the same factory. SMAW and CO2-shielded GMAW were used. Breathing zone dust levels were under 10 mg/m3 in eighty eight percent of the welding areas. Chest X-rays indicated that twenty five percent of the gas metal arc welders had siderosis (iron deposits in the lung) which he concluded was a benign disease. Bronchitis was more frequent in welders, and a synergistic effect between smoking and welding was apparent. A reduction in FEV was noted among welders and appeared to be related to the duration of welding exposure.

Schellhas (Ref. 136) examined the health status of sixteen grinders and seventeen stainless steel welders performing argon-shielded GTAW using filler materials containing nine percent nickel and nineteen percent chromium. Urinary chromium and nickel concentrations were equal to or less than that of a nonexposed population and were substantially lower than maximum acceptable values. Breathing zone levels of chromium and nickel were also well below permissible levels. Pulmonary function test results were normal, and no health effects were attributable to chromium or nickel exposure in any of the workers. These results were attributed to the excellent local exhaust and general ventilation at the plant studied.

In 1982, Zober reported that gas tungsten arc welders had a notably lower incidence of respiratory tract abnormalities than welders who used other techniques. This was presumably related to the low concentrations of fumes produced by GTAW (Ref. 167).

A follow-up study by Zober et al. (Ref. 171) examined a group of ten welders with an average welding experience of twenty years. Of the metal welded by these workers, seventy percent was stainless steel containing eighteen percent chromium and ten percent nickel. Exposures to welding emissions tended to be low. The average breathing zone concentration of total solids was 2.7 mg/m3, while that of nickel was 2.3 ug/m3, and the value for chromium was 9.8 ug/m3. Hexavalent chromium, in concentrations of 0.1 and 0.3 ug/m3 respectively, was only detected in breathing zone samples collected from two welders. The mean concentrations of chromium and nickel in all urine samples were well within normal limits. Nickel and chromium levels did not change during the work shift.

Chronic bronchitis occurred only among heavy smokers. Examination of the upper respiratory tract indicated no work-related inflammation or lesions. Lung function tests and chest sounds were normal. Shadows and irregularities were seen in chest X-rays of five of the welders. Three of these cases were persons who, previous to this study, had worked extensively with welding methods that generate considerably more fumes than GTAW. Zober et al. concluded that no clinical or medical findings could be related to GTAW of stainless steel.

In another study, Zober and Weltle (Ref. 172) examined respiratory effects of arc welding among 305 welders from twelve different companies who had an average of twenty one years experience with SMAW and GMAW of mild and stainless steel. Breathing zone concentrations were measured in seven of the twelve plants. Airborne levels of total dust, nickel, zinc, lead, manganese, copper, and chromium, but not ozone, nitrogen oxides, or hydrogen fluoride, had occasional excursions above acceptable limits, especially in confined spaces. Acceptable urine levels of chromium, nickel and cadmium were exceeded in seventy six percent, forty seven percent, and ten percent, respectively, of the samples.

An excess of bronchitis was related to smoking rather than welding, and there was no excess of pneumonia among welders. FVC and FEV_1 were significantly lower in welders than controls. The most notable changes in pulmonary function were observed in non-smoking welders above forty three years of age, which correlated with the duration of welding experience. The filler metal, welding procedure and work in confined spaces did not influence the results of pulmonary function tests. Small round opacities characteristic of siderosis were twenty seven percent more frequent in chest X-rays of welders than in controls. The incidence of these opacities was greatest in CO2-shielded gas metal arc welders and was lowest in gas tungsten arc welders. There was no relationship between pulmonary fibrosis and welding or changes in pulmonary function and the incidence of shadows in chest X-rays.

Barnhart et al. (Ref. 15) compared respiratory symptoms between 113 welding fume-exposed pipe fitters and 86 refrigeration mechanics with minimal-to-no history of exposure to welding fumes. Despite being younger and having worked fewer years in the trade, pipefitters had a significantly greater prevalence of chronic bronchitis than did the refrigeration mechanics. Differences in pulmonary function (FVC and FEV_1) were not observed except after inhalation challenges with isoetharine mesylate, a bronchodilator. The investigators suggested that this change in bronchodilator responsiveness indicates that welders may have an increased risk for bronchial hyperreactivity.

Radiographic chest lesions were studied in welders from the Dalian shipyard in China (Ref. 88). In 1960, chest X-ray revealed signs of pneumoconiosis in 28.5 percent of 400 welders. Of these, 204 persons who had worked in the shipyard for at least seven years were selected for follow-up study over the next twenty two years. During this time, three died; two from complications of extensive pneumoconiosis and the third from lung cancer. The number of cases of pneumoconiosis increased from fifty observed in 1960 to ninety at the conclusion of the study. Ten participants with pneumoconiosis were removed from dust exposure within three years after the study began. With time, the disease progressed from Stage II to Stage III in three of these persons. The FEV was significantly lower in persons with pneumoconiosis whereas there was no difference in the vital capacity or the maximum voluntary ventilation.

Lung function, clinical symptoms and chest X-rays were examined at five-year-intervals for twenty years in a group of 216 shipyard welders (Ref. 85). At the end of the twenty-year study period, 55 welders remained in the study population. Work conditions were less than standard, and dust concentrations between 20 and 95 mg/m3 were measured at various times during the study. Chronic bronchitis was diagnosed in forty to sixty one percent of welders throughout the course of the study. However the number of welders who smoked (forty four to seventy five percent) was also very high.

Pneumoconiosis developed early in the career of welders, but tended to be minimal and without concurrent abnormalities in lung function. After fifteen years of exposure, thirty three of seventy six had signs of pneumoconiosis. During the next five years, radiologic lesions progressed in seven welders (twenty one percent) and regressed in thirteen (thirty nine percent). Total regression was seen in a small number of welders. Of lung function tests performed in 1976, FEV_1 was significantly lower, but functional residual capacity and total resistance were normal. The authors concluded that the most serious lung changes observed in this welding population was bronchitis. However, the influence of smoking on the development of bronchitis cannot be excluded since seventy five percent of the study population were smokers during the first decade of the study.

Reichel (Refs. 130 and 132) surveyed the health status of the respiratory tract in 166 welders employed at three steel engineering plants. No statistically significant differences in the prevalence of cough or pulmonary function were seen between welders and controls. X-rays revealed minimal changes that could not be positively characterized as pneumoconiosis. These changes were seen in all study participants, but radio-opaque shadows were more frequent in welders than in controls. Welders who were heavy smokers had the greatest increase in the frequency of lung opacities. No signs of massive fibrosis or progressive pulmonary fibrosis were seen in either group. Based on this investigation as well as a literature survey (Ref. 131), Reichel concluded that the fibrosis associated with the small X-ray opacities seen in arc welders lungs do not cause changes in the function or integrity of the lung and therefore should not be considered to be a disease condition.

Stanulla and Liebetrau (Ref. 144) examined lung biopsy material from thirty six welders who had been referred to a heart and lung clinic in East Germany. Mild dyspnea and chronic bronchitis were found in seventy one percent and fifty two percent of the patients, respectively. Small opacities with slight fibrosis were evident in chest X-rays of twenty eight and extensive fibrosis was detected in four individuals. The extent of siderosis and fibrosis did not correlate with years of exposure. Because there were only minimal changes in lung function and clinical lung symptoms, these investigators concluded that the observed "siderofibrosis" characteristic of arc welders lungs cannot be considered to be a disease state.

To assure that epidemiologic studies of occupational groups are unbiased, persons forced to leave the working population as advanced age or decreased health status develops must be considered. A Japanese study of lung function among dust-exposed workers demonstrated that workers remaining in the area of exposure are typical healthy survivors or persons whose disease state has not progressed to the point where it is disabling (Refs. 66 and 44). An intensive effort is necessary to trace dust-exposed workers who are no longer on the job to obtain vital statistics about the entire workforce. Thus, in epidemiologic studies of welders, a lack of correlation between age and abnormalities in lung function may, in part, reflect the disappearance from the work force of diseased or older workers, or both.

In a study of 486,000 male death records of nine occupations in the State of Washington (see below), Milham reported excess mortality from bronchitis with emphysema (fourteen deaths observed, six expected) and chronic interstitial pneumonia (nine deaths observed, five expected) among welders (Ref. 92).

In summary, half of the twelve studies cited above in which pulmonary function was examined, found no significant abnormalities in the results of these tests (Refs. 85, 130, 136, 139, 142, and 171). Mao (Ref. 88) reported an association between pneumoconiosis and abnormal pulmonary function tests, but a negative correlation was found by three other investigators (Refs. 85, 142 and 172). In two studies (Refs. 139 and 135), the absence of effects on pulmonary function among welders was attributed to good ventilation in the work area. A third study (Ref. 171) attributed the overall good health status and lack of pulmonary function abnormalties to the use of GTAW which generates little fume compared with most other commonly used welding methods. Mur et al. (Refs. 105 and 106) found that shielded metal arc welders had significant reductions in pulmonary function compared with gas metal arc welders. In addition, in one plant, welders who worked in confined space had reduced lung function compared with those in well ventilated areas. Although the results are not consistent among studies, this research suggests that welding may affect some parameters of pulmonary function, and that fume emission rates and ventilation in the work area are influential factors.

Bronchitis was found in welders in nine studies. Zober et al. (Refs. 171 and 172) attributed bronchitis exclusively to tobacco use. A substantially elevated frequency of bronchitis among welders was also found by Mal'ik et al. (Ref. 85). However, no attempt was made to account for the high rate of tobacco use among the welding cohort. A synergism between smoking and welding in the induction of bronchitis was suggested by Schneider (Ref. 137) which is consistent with previous reports (see past volumes of the *Effects of Welding on Health*). Mur et al. (Ref. 105 and 106) observed bronchitis more frequently in shielded metal arc welders than in gas metal arc welders, which implies that bronchitis may be related to the extent of exposure. Whether or not welding alone produces bronchitis remains unclear.

8. Alveolar Macrophages

Macrophages increase in number in the lungs in response to inhalation of irritant gases or particulates. Gullvag et al. (Ref. 55) examined sputum samples collected from aluminum plant workers and found that the numbers of macrophages present in the samples were related both to occupational exposures and smoking.

Bariffi (Ref. 13) obtained alveolar macrophages by bronchiolar lavage from persons with pneumoconiosis caused by exposure to asbestos, silica and welding fumes. The particulate content of macrophages reflected the occupational exposure of the subject from whom they were obtained. He concluded that the presence of specific minerals in macrophages in bronchiolar lavage fluids can be used to confirm diagnosis of dust-related pneumoconiosis as indicated by X-rays, pulmonary function tests, and clinical examination.

9. Cancer

Whether or not welders have an increased risk for lung cancer remains uncertain. Results of cancer epidemiology studies of welders have been inconsistent, but several suggested the incidence of lung cancer may be elevated. Potential human carcinogens, in the form of nickel and hexavalent chromium, may be present in significant quantities in stainless steel welding fumes. Past studies have neither refuted nor supported the hypothesis that the lung cancer risk is elevated in stainless steel welders (Ref. 145). In this regard, it has been strongly suggested (Ref. 80) that available research funds be focused on the population of welders (e.g. welders of stainless steel and nickel plated mild steel) who, on the basis of current knowledge, might have the greatest risk for developing lung cancer.

In 1977, NIOSH (Ref. 113) concluded that, even though the evidence was limited, all inorganic nickel compounds should be considered to be carcinogens. However, they indicated that the recommended standards would be considered for revision should future studies demonstrate that some nickel compounds are not carcinogenic. In light of this, Mastromatteo (Ref. 90) reviewed the epidemiologic studies of nickel published since 1977. He found that studies of nickel exposures in welders were confounded by exposures to other potential carcinogens, such as hexavalent chromium or asbestos. Mastromatteo concluded that there is insufficient evidence to consider nickel to be a cause of respiratory cancer in nickel welders. In addition, he found that studies of nickel exposures among workers in the nickel-producing and nickel-using industries showed no clear association between respiratory cancer and exposure to nickel.

Stokinger (Ref. 147) reviewed published epidemiologic and animal studies of the carcinogenicity of iron oxide (including hematiate (Fe2O3), magnetite (Fe3O4), and the gamma form of ferric oxide (Fe2O4). Investigations which indicated an association between cancer and iron oxide in welding fumes did not take concurrent exposures, such as radiation, asbestos, and smoking, into consideration. In light of more recent epidemiologic and animal studies, in which exposures to iron dusts and fumes did not cause cancer, he concluded that iron oxides are not carcinogenic.

Langard and Stern (Ref. 80) reviewed twenty one cancer epidemiology studies of welders published in the international literature before 1984. Only five of these studies showed a significantly elevated cancer risk among welders, and only one examined a cohort of stainless steel welders. They stressed, as did Zober in an earlier critique (Ref. 169), that the variations in the design of epidemiology studies and insufficient background information (e.g, data on tobacco smoking or asbestos exposure) make it difficult to compare data between studies. To enable inter-study comparisons, they reemphasized earlier recommendations that a standard protocol, to be uniformly applied to studies of welders, be developed by a central scientific body.

Two epidemiologic studies of nickel- and chromium-exposed welders were published in 1985. Gerin et al. (Ref. 45) used the case-referent approach to examine the relationship between lung cancer and nickel exposure. This approach allows data to be obtained directly from participants in the study and eliminates the bias and misinformation which may be interjected when data about deceased persons is obtained from third parties. Interviews were conducted with all living male cancer patients who could be identified throughout the city of Montreal. A total of 246 lung cancer cases were found among the 1343 cancer patients participating in the study. Of these, 29 had been exposed to nickel. Persons with nickel exposure exhibited a threefold increase in lung cancer, while there was no statistically significant association between nickel exposure and the risk of cancer development in other organs.

The group with the lowest nickel exposure had a lower risk for developing lung cancer than did those with medium or high exposures. There were no differences in the lung cancer risk associated with nickel dust, nickel fumes, or stainless steel dust. Of the occupations with nickel exposure, welding had the most "remarkable association" with lung cancer. However, an elevated lung cancer risk was also found among workers in other nickel-exposed occupations. Welders without nickel exposure had little or no increased lung cancer risk.

The association between nickel exposure and lung cancer could not be successfully investigated in this study because all of the twenty nine lung cancer cases associated with nickel exposure also had chromium exposure. When the cancer site and incidence was analyzed with regard to chromium exposure, the results were similar to nickel exposure, though the risk was somewhat lower. Gerin et al. acknowledged that the observed association between lung cancer and nickel could equally have been due to chromium or to both chromium and nickel. Although the objective to separate nickel from other exposures was not achieved, this study represents an important contribution to the analysis of risk associated with combined exposures to nickel and chromium.

A retrospective epidemiologic study of the cancer risk from exposure to nickel and chromium in welders was conducted by Becker et al. (Ref. 18). The study group consisted of 1221 welders employed in twenty five German factories whose cancer rates were compared with those of 1694 machinists who worked in the same factories as well as with the general male population in the Federal Republic of Germany. Mortality statistics were collected from death certificates. Seventy seven welders and 163 machinists died during the study period. The cancer mortality rate was significantly increased in welders. Two mesotheliomas occurred in the welding group, but none was found in the machinists. The total number of deaths available for analysis at the time of this report was too small to allow statistical evaluation of the incidence of cancer in specific organs.

Milham (Ref. 92) examined the number of deaths from cancer among the death records of 486,000 adult men filed in the State of Washington between 1950 and 1982. Welders were among nine occupations included in the study. No significant increases in cancer were observed among welders.

Newhouse et al. (Ref. 110) calculated mortality rates of 1027 welders, 235 caulkers, 557 platers and 1670 electricians who worked in a British shipyard between 1940 and 1968. About fifteen percent had died by the end of the study period. The number of deaths from all causes was slightly, but significantly, higher than that expected for the welders. There were thirteen deaths from mesotheliomas; nine among electricians, two among platers, and one each among caulkers and welders. Welders and caulkers experienced a nonsignificant increase in mortality from pneumonia and lung cancer. When the rates for the welders and caulkers were combined, the increase in deaths from lung cancer became statistically significant.

Esnault et al. (Ref. 38) examined the mortality and causes of death among 100 welders who worked at a French shipyard for a period of sixteen years. Ventilation was poor during much of this time. Twenty welders died during the study period compared with thirty eight expected deaths. An excess of deaths from hepatic cirrhosis, prostate cancer, and cancer of the larynx were observed, but the differences were not statistically significant. The standard mortality ratio (calculated by comparing the number of actual deaths from specific causes with those expected on the basis of death rates in sex- and age-matched controls) for lung cancer was 0.8.

Spacilova and Hykes (Ref. 142) noted that of eleven deceased welders who had been hospitalized at the Clinic for Occupational Diseases in Prague, four died from lung cancer and three from cancer of other organs. The eleven deceased welders were part of a total group of eighty seven welders who were treated at this clinic. Although the authors recognized that these numbers were too small to be conclusive, they contended that these cases might lend support to the hypothesis that welders have an increased risk for lung cancer. However, the presumed association between lung cancer and welding is rather weak since all four welders who died from lung cancer were habitual cigarette smokers.

Exposures incidental to the welding process may increase the welder's risk of developing cancer. The now classic example of this is the development of lung cancer and mesothelioma in asbestos-exposed shipyard welders. A possible combined effect of exposure to tars and welding emissions was investigated by the United Automobile Workers Union in their epidemiologic investigation of the cause of death of ninety one millwrights and welders employed between 1966 and 1982 in a metal stamping plant (Ref. 138). The work area was poorly ventilated during the time of the study period. The floor in the plant consisted of wood blocks set in and protected with coal tar. Millwrights were exposed to tars and vapors while laying the floor tar. Both millwrights and maintenance welders were exposed to emissions produced while welding or flame cutting floor-bolts.

The twenty two observed cancer deaths were nearly twice the 14.82 expected (PMR = 189). Two-to-fivefold excesses in the number of deaths from cancer of the lung, digestive organs, testes, and leukemia accounted for most of the excess cancer cases among the millwrights and welders. Ambient air measurements showed an excessive exposure to carcinogenic polycyclic aromatic hydrocarbons, which most likely derived from the coal tar used on the floors. The investigators concluded that the excess cancer deaths may have been related to the airborne polycyclic aromatic hydrocarbons as well as to welding emissions.

Of the six cancer studies discussed above, three showed a significant excess of cancer in welders. Silverstein et al. (Ref. 138) found an excess of cancer of the lung, digestive organs, testes, and leukemia in persons exposed to welding fumes and polycyclic aromatic hydrocarbons. A significant excess of lung cancer in welders exposed to nickel and chromium was found by Gerin, (Ref. 54) and an increase in the overall cancer rate in stainless steel welders was demonstrated by Becker et al. (Ref. 18).

Mortality studies performed by Milham (Ref. 92), Newhouse (Ref. 110), and Esnault (Ref. 38) did not indicate an elevated cancer rate among welders. The latter studies investigated death rates of welders as a group and did not separate out cases on the basis of specific fume components. Although none of these studies allow the firm conclusion that welding fumes are carcinogenic, they do support the premise of Langard and Stern (Ref. 80) that epidemiologic studies should focus on populations of welders exposed to fumes containing known or suspected carcinogenic components.

10. Metal Fume Fever

Metal fume fever, a common occupational illness among welders, is caused by inhalation of metal oxides including zinc, copper, aluminum, antimony, iron, manganese, nickel, and cadmium. Fever, chills, general malaise, joint pains, cough, sore throat, chest tightness, and fatigue usually appear four to twelve hours following exposure and last from one to two days. Symptoms may reappear after several days away from work; hence, workers frequently refer to it as "Monday fever". Diagnosis of metal fume fever is sometimes difficult because the symptoms resemble those of a number of upper respiratory tract illnesses.

The early symptoms of acute exposure to cadmium fumes may resemble metal fume fever, but the clinical course of the disease can differ markedly. Severe cadmium poisoning can cause chemical pneumonitis with extreme breathing difficulties, cough, wheezing, abdominal pain, and headache. Kidney failure, pulmonary edema, and respiratory failure may develop in cases of severe poisoning. Chronic exposure to cadmium may cause emphysema, pulmonary fibrosis, renal insufficiency, and possibly cancer.

Because of the similarities between metal fume fever and the initial stages of acute cadmium poisoning, Barnhart and Rosenstock (Ref. 14) stressed that possible exposure to cadmium fumes should be considered whenever patients show symptoms of metal fume fever. They reported the case of a welder who, after working with silver solder in an enclosed area for one hour, developed shortness of breath, cough, fever, and muscle pain. He was diagnosed as having metal fume fever although the cough and shortness of breath persisted for four weeks. Because of the potential seriousness of cadmium poisoning, the authors concluded that the "diagnosis of chemical pneumonitis secondary to cadmium fume exposure should be suspected in any patient complaining of shortness of breath following exposure to metal fumes."

The pathogenesis of metal fume fever is poorly understood. An immune reaction to metals, resulting in inflammation of respiratory tract tissue and release of histamine or histamine-like substances, is the most widely accepted theory concerning the etiology of this malady (Ref. 104). Pantucek (Refs. 124 and 125) attempted to explore mechanisms of metal fume fever by measuring body temperature following exposure to zinc oxide through spot welding and open flame welding of galvanized steel, and galvanizing small objects in an open zinc bath. Fever was not produced by exposure to any of these processes. However, zinc levels in urine or serum were not elevated in any of the workers. Thus, in this investigation, the zinc levels to which workers were exposed may have been too low to adequately test whether zinc oxide fumes can produce elevations of body temperature in the absence of other symptoms of metal fume fever.

11. Effects on the Ear and Hearing

Most forms of welding, and particularly plasma arc cutting, can generate noise in excess of permissible levels. Welders may suffer hearing loss as a result of exposure to high noise levels generated by welding and by other operations carried out in their vicinity.

Studies of the effects of occupational noise performed under the auspices of the Academy of Finland indicate that impulse noise may be more harmful than continuous noise. As part of this project, Mantysalo and Vuori (Ref. 87) compared hearing thresholds (minimum detectable sound levels) in shipyard workers exposed to impulse noise and in workers from a cable factory exposed to continuous steady state noise.

In this study, damage to hearing occurred after shorter exposures to impulse noise than to continuous noise. Exposure to impulse noise for three to four years produced effects similar to those resulting from exposure to an equivalent level of continuous noise for five years. The frequencies most sensitive to impulse noise were between 4000 and 6000 Hz. The longer the duration of exposure to impulse noise, the wider the range of frequencies which showed raised thresholds. Because of the differences in effects on hearing, the investigators noted that hearing protectors may require different qualities for protection against impulse noise than for continuous noise.

12. Effects on the Eye and Vision

Arc welding generates electromagnetic radiation in the infrared (IR), visible and ultraviolet (UV) range, all of which can affect the eye. Short intense exposure to visible and near IR radiation may injure the retina. Infrared radiation of longer wavelengths may cause thermal damage to the cornea and aqueous humor and has also been associated with the development of lenticular cataracts. Short-wave UV light (270 to 290 nm) is absorbed by the outer layers of the cornea and can cause photokeratitis (arc eye, welder's flash, photophthalmia); long-wave UV light reaches the inner layer of the cornea and the lens. Possible effects of UV light on the corneal endothelium (Ref. 72) and the lens (Ref. 84) are discussed below.

Photokeratitis, a marked inflammation of the cornea, is the most common eye problem encountered by welders. The symptoms, which include blurred vision, tearing, acute pain, and headache, may last up to two days and usually have no sequelae. The symptoms of photokeratitis are preceded by a latent period of four to five hours during which time there is no pain. To investigate reasons for the lack of symptoms during this characteristic, long latent period, Millodot and Earlam (Ref. 93) measured the sensitivity of the cornea to touch during the first five hours after exposure. Seven volunteers were subjected to the arc from a portable electric-arc welding set. One eye was kept closed during the three second exposure and served as a control.

The corneal touch threshold (CTT—the lowest pressure on the eye that can be felt by the subject) increased an average of seventy three percent in the exposed eye. It peaked at 1.75 hours and returned to normal by 4.25 hours after exposure (Figure 4). The period of diminished sensitivity corresponded to the latent period between exposure to UV radiation and appearance of symptoms of photokeratitis. The authors postulated that this reduced sensitivity is responsible for the lack of symptoms during the first few hours after exposure to the welding arc. An alternative explanation is that the altered sensitivity to pain is part of a series of complex biochemical and physiological responses to UV exposure which eventually lead to the appearance of symptoms of photokeratitis.

Prolonged exposure to reflected sunlight at high altitudes can cause photokeratitis. Blumthaler et al. (Ref. 23) calculated that the total dose of reflected sunlight required to produce photokeratitis is substantially higher than that from the electric welding arc. He reasoned that the threshold dose depends on the intensity of the radiation and increases with decreasing intensity. With a less intense UV light source, such as reflected sunlight, repair processes may have more time to operate, providing some measure of protection during exposure.

Among other eye lesions thought to be caused by ultraviolet radiation are pterygia, membranous growths which extend across the outer eye from the conjunctiva to the cornea. To determine if this lesion can be caused by exposure to the welding arc, Karai and Horiguchi (Ref. 71) studied the incidence of pterygia in welders. Slit lamp examinations of the

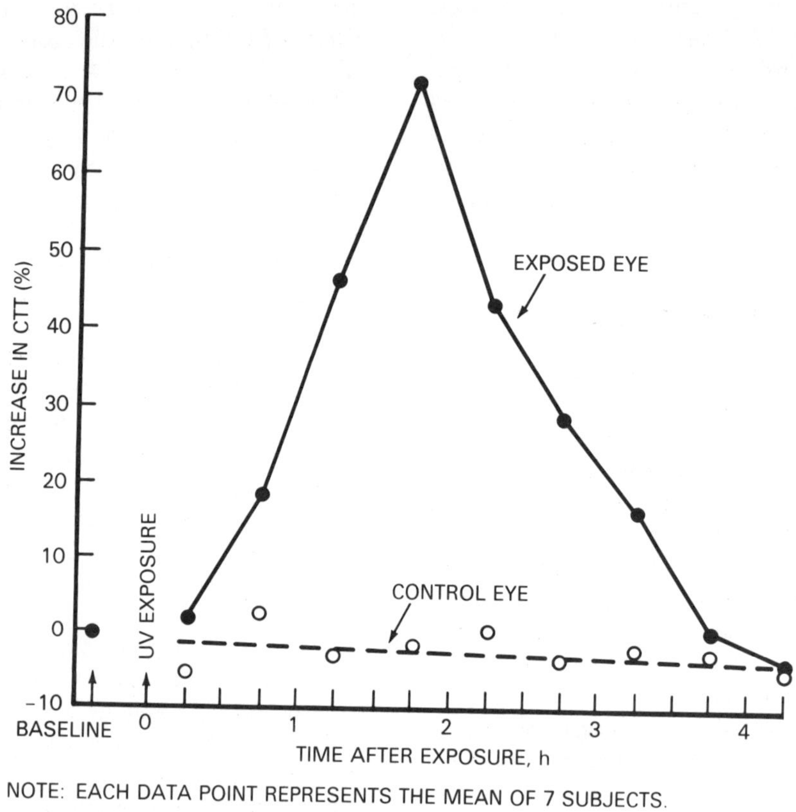

Redrawn from Millodot and Earlam (Ref. 93)
Note: Each data point represents the mean of 7 subjects

Figure 4 — Percent Increase in CTT as a Function of Time After Exposure to UV Light

conjunctiva and corneas of 191 welders from three Japanese factories were compared to 214 nonwelders who worked in the same plants. Pterygia were found in 17 welders and only one worker from the control group ($p<0.001$). In addition, there was a significant relationship between the duration of employment as a welder and the incidence of the lesion. Karai and Horiguchi concluded that there is a significant relationship between the incidence of pterygia and exposure to UV radiation in welders.

The corneal endothelium is a monolayer of multi-sided cells covering the posterior surface of the cornea. The activity of these cells is essential to maintenance of the thickness of the cornea. With age, the cells enlarge and become more multi-sided, while there is a decrease in the total cell number. These changes with age have no known effects on corneal function.

The possible role of UV light on these cellular alterations in the corneal endothelium was examined by Karai et al. (Ref. 72) in 118 welders and 85 nonwelders. There were no differences in mean cell size between the two groups, and the only significant difference was a decrease in the number of hexagonal cells in the welding population. Although Karai concluded that UV light is involved in the aging of the endothelium, the data did not indicate that welding has deleterious effects on the function of the corneal endothelium.

The relationship between short-wave IR radiation and cataract formation was investigated by Lydahl et al. (Ref. 84) who examined the eyes of 208 IR-exposed workers from six iron and steel manufacturing plants. An increased prevalence of wedge-shaped lenticular opacities, considered to be senile cataracts, was found in the lenses of IR-exposed workers over fifty nine years old as compared to age-matched controls. Lydahl concluded that exposure to IR radiation may accelerate changes in the lens that normally occur with aging.

Conflicting results were obtained by Dvorak et al. (Ref. 34) who examined the eyes of thirty eight shielded metal arc welders to investigate whether welding exposures cause chronic or permanent damage to the eye. No permanent deterioration of the lens specifically attributable to welding was found.

The effect of welding on the function of the retina was examined by Dobromyslov et al. (Ref. 31) using the photo-stress test. This method measures the time required to restore visual acuity following illumination of the retina. After visual acuity was measured under normal conditions, one eye was illuminated with an electro-ophthalmoscope for thirty seconds. The time required for restoration of initial visual acuity (TRIVA) was measured separately in each eye of persons with over ten years welding experience and in controls. The welders in this study used either argon-shielded GTAW or CO2-shielded GMAW of stainless steel. The TRIVA was longer in welders than in controls (Table 5). This difference was greatest in GMAW welders thirty one to forty years of age. With the exception of a small decrease in the field of peripheral vision, no other welding-related eye abnormalities were observed.

A glare test, similar to the photo-stress test used by Dobromyslov, was included among other vision tests administered by Gos et al. (Refs. 51 and 52) to seventy four welders and forty office employees from a plant in Poland. No significant differences in stereoscopic vision or in the TRIVA were observed between welders and controls. Differences in the results of the glare tests obtained by Dobromyslov and Gos et al. are not explainable with the available data. Gos et al. did not address the types of welding methods used by their subjects. Hence, it is possible that differences in results may be related to the different radiation spectra produced by different welding procedures.

Table 5
Effect on Welding on Time Required for Restoration of Initial Visual Acuity (Triva) Following Exposure to Light

Subject	Number Persons	Age (Years)	TRIVA (Seconds)
Control	34	26–40	67.6 ± 1.09
	NG	26–30	67.4 ± 1.14
	NG	31–40	68.4 ± 1.05
Welders			
GTAW	22	31–40	75.6 ± 7.44
GMAW	8	26–30	76.9 ± 4.6
GMAW	32	31–40	100.8 ± 4.47

NG—data not given
Data from Dobromyslov et al., Ref. 31

Gos et al. (Ref. 52) observed a statistically significant increase in the ability of welders to differentiate shapes at minimum illumination. The investigators surmised that the welders' superior performance in the latter test may have resulted from their continual adaptation to rapid changes of illumination during welding. Degenerative changes in the retina and symptoms of conjunctivitis were found more frequently in welders than in controls.

During the last fifteen years, there have been several reports in the popular press of accidents resulting from persons wearing contact lenses while welding. Presumably, the radiation emitted by the electric arc caused the lacrimal film covering the surface of the eye to evaporate resulting in adhesion of the cornea to the contact lens and loss of vision. Considering physical principles involving the radiant spectra generated by the electric welding arc, the behavior of electromagnetic radiation, and the absorption of electromagnetic radiation by water, biological tissues and synthetic materials, Preusz and Geyer (Ref. 128) demonstrated that sufficient radiation could not physically reach the eye from the electric arc to cause this effect. They concluded that the dangers of radiation to the eye are the same for persons who wear contact lenses as for those who do not.

13. Skin

Skin lesions experienced by welders include allergic dermatitis, resulting from exposures to materials such as chromium compounds, and ultraviolet-induced erythema (sunburn). Burns can result from flying sparks, and less frequently from molten slag or contact with open flame. Eun et al. (Ref. 39) reported that 8.8% of 4325 industrial workers examined in South Korea in 1981 had signs of skin burns. Most of the skin burns observed among welders were first or second degree burns and few required hospitalization.

14. Sensitivity to Fume Components

Allergic contact sensitivity to chromium among welders is well documented. In 1985, four cases of allergic dermatitis were reported by Morris et al. (Ref. 102). Four stainless steel welders developed skin lesions and other symptoms following a period of frequent and intensive use of high-chromium welding rods. They developed itchy rashes on the ankles, hands and forearms, severely fissured lip lesions, bloody nasal discharge, muscle and joint pain, and inflammation of the esophagus. Plasma and urinary chromium were elevated at the time of the initial examination. The chromium levels returned to normal and clinical symptoms improved after they stopped using high-chromium electrodes. Plasma and urinary chromium were normal in a fifth, asymptomatic welder who had used high-chromium electrodes for six years.

15. Effects on the Cardiovascular System

The use of respirators can place physiological and psychological strain on the worker. The most important factors that contribute to this are the weight of the apparatus and the inspiratory or expiratory breathing resistance, or both. Louhevaara et al. (Ref. 82) investigated the burden imposed by the regular use of respirators on the cardiovascular and respiratory systems. The effects of industrial respirators on the heart rate, oxygen consumption and ventilation rate of twenty one workers were measured. Filtering devices, air-line (supplied-air) devices, and self-contained breathing apparatuses were studied. Of these, the filtering devices used by welders and other workers caused hypoventilation, retention of carbon dioxide, and elevated heart rates. Supplied-air devices had smaller effects on heart rate and gas exchange and hence were recommended for use in preference to air filtering devices. The investigators further recommended that the continuous use of respirators be limited to the shortest periods possible (less than thirty minutes) and that proper rest pauses be taken.

16. Effects on the Nervous System

Anatovskaya (Ref. 7) surveyed the neurological status of 180 persons suffering from chronic bronchitis (54 welders, 92 foundry workers, and 34 grinders) at an occupational health clinic in Russia. Similar neurological symptoms were seen among the three occupational groups. These included weakness, exhaustion, fatigue, apathy, headache, dizziness, imbalance, numbness or pain in the extremities, irritability and memory loss. The degree and frequency of nervous system changes increased with the severity of the pulmonary insufficiency and were thought to have resulted from inadequate oxygen supply to the brain.

17. Effects on the Musculoskeletal System

Muscular strain in industrial work can cause a variety of pathologic processes and work disability. The underlying pathophysiology of chronic shoulder pain is poorly understood. In 1984, Herberts et al. (Ref. 60) published a study of the prevalence of shoulder pain in shipyard platers and welders which was largely a recapitulation of material published in 1981 (Ref. 59). Briefly, their working hypothesis was that specific working postures induce heavy shoulder muscle load which increased the risk for sustaining chronic shoulder pain and disability. They found that shoulder pain, or more specifically supraspinatus tendonitis, was significantly more prevalent in welders and platers than in office workers. Supraspinatus tendonitis was equally common in the welders and platers but it appeared at an earlier age in welders. The most important difference between the two occupations is that welding can be characterized as almost exclusively static whereas more dynamic movements are used by platers. Thus, static work may entail a higher risk for chronic shoulder pain. Stone (Ref. 148) also reported that the most common repetitive strain injury among welders is caused by static load strain.

18. Effects on the Urogenital Tract

Chronic low-level exposure to cadmium fumes can lead to kidney damage which is first manifested by urinary excretion of low molecular weight serum proteins (proteinuria) such as beta-2-microglobulin. The incidence of proteinuria increases with years of exposure to cadmium. It may be present with little other evidence of kidney damage. Previous studies and those described below (Refs. 35 and 36) indicate that measurement of levels of urinary beta-2-microglobulin or other low molecular weight proteins is a useful tool for identifying early adverse health effects and for prevention of the more serious kidney damage that can be caused by cadmium exposure.

Chiesura et al. (Ref. 24) examined urine specimens from sixteen persons in four workshops where brazing was performed with high-cadmium alloys. Ambient cadmium levels in the brazing areas ranged from 0.15 to 1.16 mg/m3. Blood and urine cadmium levels were related to the duration and severity of exposure. Urinary beta-2-microglobulin was high in two of the workers while other proteins, also indicative of tubular dysfunction, were found in urine samples from other workers. One case of nephropathy was attributed to cadmium exposure (Ref. 159).

Elinder et al. (Refs. 35 and 36) characterized the cadmium-induced renal effects in sixty workers who were exposed to cadmium-containing solders for four to twenty four years at a factory which produced radiators and heat exchangers. The cumulative dose received over the entire exposure period was estimated for each participant and ranged from 0.35 to 9.9 mg/m3 per year. Slightly elevated urinary beta-2-microglobulin levels were seen in forty percent, and more pronounced levels occurred in twenty three percent, of the participants. Urinary beta-2-microglobulin levels correlated well with urinary cadmium and with the estimated cumulative cadmium dose. Among cadmium-exposed workers, those with a history of kidney stones had a significantly higher urinary cadmium concentration and tended to have higher levels of urinary beta-2-microglobulin than those with no history of stones.

Urinary beta-2-microglobulin levels were followed in nineteen of the workers over an eight year period. After the first two years, the factory was cleaned up and renovated. Cadmium was not used during the remaining six years of the study period. Urinary beta-2-microglobulin remained elevated in most of the subjects even six years after exposure had ceased, suggesting that cadmium-induced proteinuria is of long duration, if not irreversible.

The exact nature of the kidney dysfunction caused by cadmium is somewhat controversial. Cadmium intoxication is generally thought to damage the kidney tubules. However, this has been challenged by some investigators who observed urinary excretion of high molecular weight proteins such as albumin, transferrin, and immunoglobulin (IgG) in persons with high cadmium exposures. The controversy revolves around the issue of whether elevations in high molecular weight proteins reflect an increased permeability of the glomeruli or a decreased resorption of high molecular weight proteins through the kidney tubules.

During normal kidney function, low molecular weight proteins are filtered through the glomeruli and resorbed through the tubules. Since only minimal quantities of high molecular weight proteins can pass through normally functioning glomeruli, an elevation of high molecular weight proteins would indicate glomerular dysfunction. If the glomeruli are functioning properly, but the tubules are not, then the low molecular weight proteins would not be resorbed by the tubules and abnormally high levels of low molecular weight proteins would appear in the urine.

Using high resolution cellulose acetate electrophoresis, Falck et al. (Ref. 40) observed both high and low molecular weight proteins in urine samples from seven cadmium-exposed workers. They concluded that both glomerular and tubular damage can result from cadmium exposure.

Elinder et al. (Ref. 36) found that urine concentrations of the low molecular weight proteins beta-2-microglobulin and orosomucoid were substantially elevated while albumin (a high molecular weight protein) was only minimally elevated in workers with high cadmium exposures. Since small amounts of high molecular weight proteins can normally pass through the glomerulus, Elinder et al. interpreted this data to indicate a decrease in tubular resorption rather than an increased glomerular permeability for larger proteins. In this way, tubular impairment could also result in the excretion of larger proteins. The question of whether only tubular function, or both tubular and glomerular function, are affected by cadmium exposure remains unresolved. Because of the value of urinary proteins in indicating the extent and type of injury that can result from cadmium exposure, this remains an important and active area of scientific investigation.

19. Effects on the Endocrine System

Smirnov et al. (Ref. 141) determined urinary levels of 17-hydroxycorticosteroids (17-OHC), adrenaline and noradrenalin as a measure of strain and physical stress. Urine samples were collected over twenty four-hour periods from healthy males working in the following occupations: university staff; thermal power station operators; electric welders with generally satisfactory work conditions; electric welders with poor work conditions; and brick work inspectors. No significant differences in total urine production or 17-OHC levels were observed. Adrenalin and noradrenalin levels were significantly higher in power plant operators and in welders working in poor conditions than in the other occupational groups. The investigators speculated that these increases were related to emotional stress in the former group and to heavier work loads and less favorable working conditions among the welders. With the exception of abnormally high levels of adrenalin and noradrenalin in thermal power station operators and welders exposed to poor work conditions, the mean levels of 17-OHC, noradrenalin and adrenalin excretion in all five groups were within physiologically normal limits.

20. Effects on the Teeth and Oral Cavity

A survey of Swedish commercial divers indicated that thirty six of sixty six who performed underwater electric arc welding experienced a metallic taste in their mouth, possibly derived from degeneration of dental amalgams, when working with electrical equipment under water. This sensation was not experienced by twenty four divers who did not work with such equipment (Ref. 122). This effect, as well as an unusually high replacement rate of dental amalgams, was noted previously (Ref. 25). Clinical examination showed that the appearance of dental amalgam restorations in twenty nine divers who performed underwater welding or cutting during the last two years differed significantly from those of eleven divers who had not worked with underwater electrical equipment during this time period. This difference could not be related to salivary secretion rate, buffering capacity, secondary caries, or cultures of *S. mutans* and *Lactobacillus* (Ref. 122).

21. Effects of Hyperbaric Pressure

In a report released by the Foundation of Scientific and Industrial Research at the Norwegian Institute of Technology, Bjorseth et al. (Refs. 20 and 21) discussed the need for research on the toxic effects of gases and fumes at great oceanic depths as underwater welding becomes more extensive with the growth of off-shore oil exploration. Divers are used for installation, welding and maintenance of equipment at depths greater than 300 meters. Under such high hydrostatic pressure, the toxic effects of chemicals may change and the exposures may alter the physiological stress already inherent in deep sea diving.

Previous studies indicated that adverse physiological effects may result from exposure to hyperbaric conditions even in the absence of chemical exposures. These effects include nervous excitation, and changes in pulmonary function and CO2 sensitivity. *In vitro* studies revealed changes in macromolecular synthesis, enzyme activity, membrane structure and function, and cell division. How these observations relate to hyperbaric effects on the intact organism is unknown.

It is quite possible that the stress already imposed upon the body by hyperbaric pressure will alter the toxic effects of chemicals. Studies of a few medications, anesthetics and breathing gases have shown that effects seen at normal atmospheric pressures cannot be extrapolated to hyperbaric conditions. Such studies are few, and the combined effects of hyperbaric pressures and exposures to industrial chemicals have not been studied.

In addition to possible modification of the chemical toxicity of welding fumes, hyperbaric pressure may also alter the composition of welding emissions. Preliminary experiments with helium-shielded GTAW of mild steel indicated that hyperbaric pressure may cause changes in the chemical composition and emission rates of welding fumes. The fume generation rate increased with pressure and was five to ten times greater at 30 bar helium than at 1 bar helium. At 30 bar pressure, the ratio of manganese to iron was reduced and both thorium and tungsten were detected in fume samples.

Effects of hyperbaric pressure on concentrations of carbon dioxide, carbon monoxide, ozone and other gases generated by welding are not known. Further research into the fumes and gases produced during underwater welding is critical to the development of safe working conditions in that environment. A hyperbaric ozone monitor is currently being developed at the Norwegian Institute of Technology for research in this area. Currently, analysis of welding emissions generated within hyperbaric chambers can only be performed by capturing gases and fumes and analyzing them outside of the chamber at normobaric pressures. Instrumentation should be developed for enabling analysis of aerosols and gases under hyperbaric conditions. More studies are needed of underwater welding emission rates and means for effective ventilation.

Currently in Norway, chemical exposure guidelines for normal atmospheric conditions are used for hyperbaric work. New research into the effects of exposure to gases and fumes under hyperbaric conditions is necessary for the development of exposure guidelines appropriate for use in underwater welding. Bjorseth et al. recommended that investigations be performed to explore the effects of hyperbaric pressure on lung function, particle deposition, lung clearance, and the organ distribution of chemicals. Methods should be developed for toxicity testing under hyperbaric conditions; *in vitro* studies of the effects of high pressure on cell systems should be expanded.

22. Biological Monitoring

22.1 Nickel and Chromium. Many studies have been performed in which urine or blood levels, or both, of nickel or chromium were analyzed to determine their utility for monitoring worker exposure. Aitio (Ref. 2) reviewed the published literature on biological monitoring of occupational exposure to nickel. He concluded that this technique has little utility for estimating the health risks associated with nickel exposure.

Several recent studies support Aitio's conclusions. Akesson and Skerfving (Ref. 4) determined nickel concentrations in urine samples collected twice weekly from eleven shielded metal arc welders of stainless steel. Tests were performed for six successive weeks during which time the welders were engaged for an average of 5.8 hours per day in high-nickel alloy welding. Although the subjects of the study were seasoned welders of high-nickel alloy steel, they had not welded nickel-containing alloys for at least four weeks before the start of the investigation. The average ambient concentration of nickel in the work area was 0.44 mg/m3 with a range of 0.07 to 1.1 mg/m3 and the airborne concentration of chromium averaged 0.1 mg/m3.

Figure 5 shows the mean urine nickel levels of the welders throughout the study period. At the start of the study, urinary nickel concentrations were significantly higher in welders (8.7 ug Ni/l) than in controls (5.1 ug Ni/l), and there was no correlation between urinary nickel levels and the number of years experience welding high-nickel alloy or stainless steel. Nickel concentrations in urine collected on Monday mornings increased erratically during the six-week study period (mean 13 ug/l) and the levels were slightly, but significantly higher on Thursday afternoons (mean 18 ug/l). There was no correlation between ambient and urinary nickel levels. The investigators concluded that urine nickel levels are of little use for biological monitoring.

Zober (Refs. 168 and 170) determined levels of nickel and chromium in the breathing zone and body fluids of 20 arc welders who worked with filler materials containing eighteen to twenty percent chromium and eight to ten percent nickel. Chromium and nickel were measured in air samples collected for two hours each day during the work shift for a period of one week and in urine and plasma collected two or three times daily for one full week and the following Monday. For the purpose of analysis, welders were divided into three groups who worked with a) predominantly GTAW and GMAW with up to 20% SMAW, (b) occasional GTAW and GMAW with 20–60% SMAW; and (c) 60–100% SMAW.

As expected, exposure to total chromium, hexavalent chromium and nickel increased with the proportion of SMAW. Urinary chromium levels tended to increase during the work shift; average values for the entire welding group increased progressively from 5.3, to 7.8, to 10.3 ug/g creatinine throughout the day. A linear relationship existed between exposure levels and post-shift concentrations in the urine and plasma for both hexavalent and total chromium (Table 6). Pre-shift urinary chromium levels increased throughout the week indicating some accumulation of chromium in the body.

No such patterns could be established for nickel which varied only slightly in the urine despite substantial increases in nickel concentrations in the fume. Median plasma chromium levels in the welders were approximately ten times higher than the corresponding value for the general population, while nickel levels were the same among welders and controls. Zober et al. concluded that workplace exposures to chromium, but not to nickel, can be monitored by urine and plasma analysis. Furthermore, since there is a close correlation between plasma and urine chromium concentrations, in most circumstances it is not necessary to determine plasma chromium levels for biological monitoring.

In his review of biological monitoring of nickel exposure, Aitio (Ref. 2) cited two reports which indicated that measurements of chromosomal aberration and sister chromatid exchange rates do not reflect exposure levels in nickel workers. However, one of these studies (Ref. 162) did show an increase in the number of chromosome gaps in nickel-exposed workers. Although, according to Aitio, the significance of such gaps is doubtful and their importance is not clearly understood, the increase in chromosome gaps in nickel-exposed workers may be an area worthy of further research.

More recently, Koshi et al. (Ref. 75 and 76) measured the frequency of sister chromatid exchanges, chromosomal aberrations, and the number of chromosomes in lymphocytes from stainless steel welders who used both GMAW and GTAW. Welders had substantially higher urinary concentrations of chromium than did controls. This is of importance since other studies indicated that the degree of sister chro-

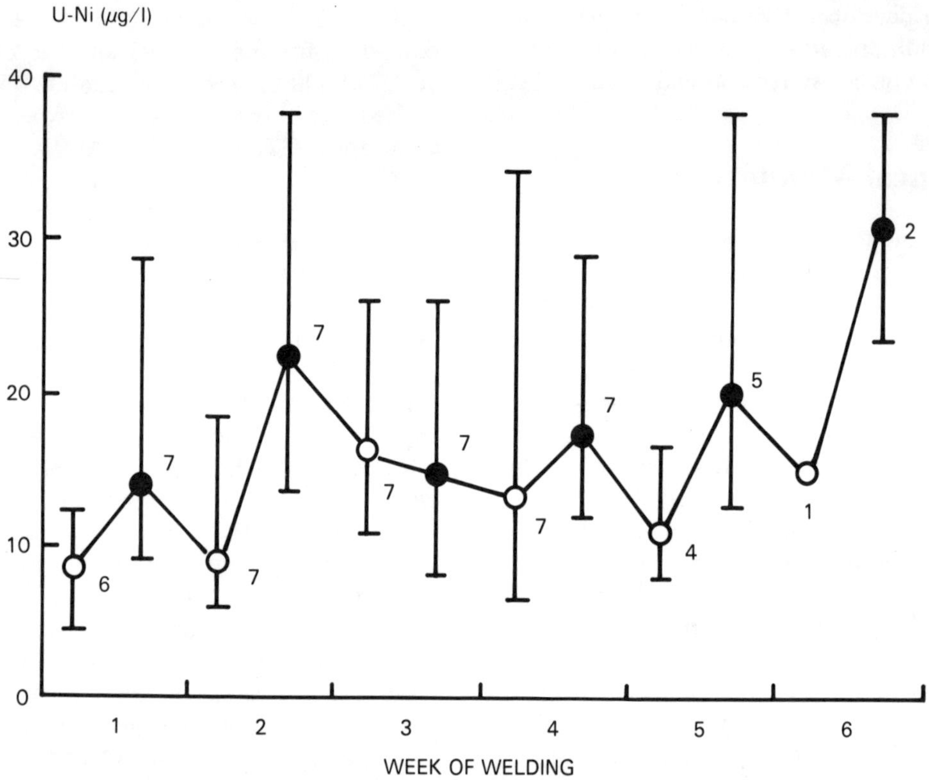

Redrawn from Akesson and Skerfuing (Ref. 4)
Note: The figures next to the symbols indicate the number of samples. The first Monday morning value was obtained after four weeks of vacation.

Figure 5 — Nickel Levels in Urine (U-Ni, mean and range) in Seven Welders on Monday Morning (Open Symbols) and Thursday Afternoon (Closed Symbols) During Six Weeks of Welding a High-Ni Alloy

Table 6
Urine and Plasma Levels of Metals and Fluoride (Median and 90% Range Exposure Groups Classified by Extent of Use of Coated Filler Metals

Compound	Group 1	Group 2	Group 3
Chromium – urine (μg/g creatinine)	2.5 (0.2–7.8)	5.7 (2.2–17.7)	19.1 (3.3–73.2)
Nickel-urine (μg/g creatinine)	1.7 (0.4–7.4)	2.4 (0.2–7.8)	3.0 (1.2–9.8)
Fluoride-urine (μg/g creatinine)	0.34 (0.18–0.85)	0.47 (0.23–0.96)	0.53 (0.2–1.92)
Chromium-plasma (μg/l)	0.8 (0.4–2.2)	2.2 (0.9–4.9)	4.9 (1.1–9.6)
Nickel-plasma (μg/l)	0.4 (0.4–1.8)	0.4 (0.4–4.9)	1.3 (0.4–5.4)

Zober, Ref. 168

matid exchange induction correlates well with the concentration of hexavalent chromium in water extracts of welding fumes (Ref. 12).

There were no significant differences in the sister chromatid exchange frequency between welders and controls. Chromosomal aberrations, including chromosome gaps, aberrant metaphases, and chromatid and chromosome gaps, occurred slightly, but significantly more frequently in welders than in controls.

22.2 Barium. Barium, an essential trace element, is toxic in high doses. Water-soluble barium compounds are muscle and nerve stimulants and excessive exposure can result in gastroenteritis, slow pulse rate, and muscle paralysis (Ref. 26). Insoluble barium is non-toxic by either inhalation or ingestion.

Dare et al. (Ref. 28) measured urine barium levels in five welders exposed for three hours to fumes generated from self-shielded flux-cored wire electrodes containing about forty percent barium carbonate in the flux. The welding fumes contained twenty five percent barium. The average urine levels were 126 ug/l (range 31–234 ug/l) immediately after exposure and 48 ug/l the following morning. Levels in three unexposed persons were under 5 ug/l.

22.3 Manganese. Like barium, manganese is an essential element which can be toxic under certain circumstances. Chronic high exposures to this metal can cause a neurological syndrome similar to Parkinson's disease. Manganese levels in plasma samples from two workers employed in cutting manganese steel dropped considerably after respirators were introduced into the work area and improvements were made in ventilation. Knight et al. (Ref. 73) recognized that the number of persons in this study and the number of samples drawn from controls were too small to enable firm statements about the value of manganese levels in biological fluids for estimating exposure risks. They concluded, however, that the reduction of manganese levels in plasma of exposed workers following reduction of fume exposure was encouraging.

22.4 Aluminum. Sjogren et al. (Ref. 140) examined the relationship between urinary aluminum levels and exposure to aluminum in fumes generated by GMAW of aluminum. Nine subjects participated in the study. Three had no previous experience with aluminum welding and were exposed to fumes containing thirty nine percent aluminum for one day. Urinary aluminum concentrations increased from 3 ug/l prior to exposure to between 15 and 414 ug/l within twenty four hours after exposure. The levels returned to pre-exposure values within a few days.

Of the remaining subjects, three had welded aluminum for less than two years and three had over eighteen years experience welding aluminum. These persons were exposed to welding fumes with an eight-hour time weighted average aluminum level of 2.4 mg/m3 for one week. Aluminum concentrations in urine from welders with short-term experience rose during the week and returned to previous exposure levels over the weekend following the five day exposure period. In those with many years welding experience, urinary aluminum levels did not change after the cessation of exposure. Sjogren et al. concluded that part of the inhaled aluminum is excreted rapidly in the urine following exposure while the remainder may be stored in the body and excreted slowly. Thus, aluminum inhaled from welding fumes may be retained for a long period of time.

22.5 Lead. Blood lead levels are generally considered to be more reliable indicators for monitoring exposure to lead than urine lead levels since the latter change slowly in response to lead exposure. Diagnostic tests which measure the effects of lead on the heme system, such as measurement of zinc erythrocyte protoporphyrin (ZEP) or urinary coproporphyrin, are considered to be useful indicators of the biological effects of lead absorption by the body, but are less useful than blood lead levels for monitoring acute lead exposures (Ref. 17 and 61).

This was exemplified in the case report of a welder who had a massive lead exposure while performing a temporary assignment in the lead oxide paste mixing plant of a lead-acid battery factory (Ref. 164). Blood lead levels of 240 and 300 ug/dl were measured on two separate occasions within two weeks after the initiation of his four-day exposure. (The normal upper limit for blood lead is 40 ug/dl.) His blood lead levels dropped steadily and returned to near normal levels within six months. Urinary lead and coproporphyrin concentrations were only slightly elevated whereas ZEP rose for eight weeks and then slowly decreaed. It is most surprising that even with the extraordinarily high blood lead levels, the welder was asymptomatic and apparently suffered no ill effects from the exposure.

Kalnas and Alleyne (Ref. 70) reported that ZEP may be a better biological screening parameter than blood lead levels for chronic lead exposure. Urine aminolevulinic acid (ALAU), blood lead (BPb), and ZEP were measured in 142 lead-exposed radiator shop workers in Alberta Canada. ZEP, but not blood lead levels reflected symptoms of lead exposure in workers employed two years or less. ALAU levels tended to be indicative of longer term exposure as ALAU, but not blood lead levels, were elevated in workers employed for more than four years.

22.6 Cadmium. A study of environmental causes of elevated blood cadmium levels in brazers was performed in Sweden (Ref. 83). The 102 brazers who participated in the the study had worked with cadmium-containing solders for at least ten percent of the work day during the preceding three months. Blood cadmium concentrations were less than 10 ug/1 (considered in this study to be the maximum acceptable blood concentration) in 71 and equal to or above 10 ug/l in thirty one workers. The material brazed, duration and extent of exposure, age, sex, and smoking habits did not influence blood cadmium levels. The single most important factor contributing to the blood cadmium elevation was the length of the splice being brazed. Blood cadmium levels greater than 10 ug/l were seen in none of the brazers working with splices shorter than 2 cm, but were observed in eighty three percent of the brazers working with splices longer than 2 cm. This value was related to the use of exhaust ventilation since local exhaust was rarely used with longer splices.

22.7 Heavy Metal Monitoring in Hair. The utility of hair analysis for monitoring exposure to heavy metals has been actively explored for many years. Hair samples are easy to collect, and the analyses can provide an integrated measure of exposure over an extended period of time. However, contamination from environmental sources can present significant problems during hair analyses. Metals in airborne dusts or vapors making direct contact with the hair can become incorporated into the hair shaft. Metals which entered the hair by direct contact are indistinguishable from metals which entered the hair after absorption by the body through other routes (e.g., inhalation or ingestion). Only the latter are of value for assessment of exposure.

A unique approach to this problem was taken by Huel et al. (Ref. 62). They examined whether cadmium and lead absorbed following occupational exposures are capable of reaching the human fetus by determining the concentrations of these metals in newborn hair samples shortly after birth. A unique feature of this study is that contamination of the hair of newborn babies by heavy metals in the environment is negligible. The twenty six experimental subjects were pregnant women whose heavy metal exposures derived mostly from soldering electronic components.

The concentrations of cadmium and lead in hair from exposed mothers and of cadmium in hair from their offspring were more than twice as high as levels present in hair from unexposed controls. Lead levels in the hair of newborns were not consistently higher in the exposed group than in the controls. It was concluded that systemic cadmium exposure can be quantified by hair analysis of either the mother or the newborn. The lack of a correlation between maternal and newborn lead levels could not be explained. The investigators suggested that higher levels in maternal hair could result from external contamination or a decreased passage of lead through the placental barrier, or both. Other factors such as differences in the distribution of the two metals in the tissues of the fetus must be considered as well.

Gorban et al. (Ref. 49) determined the manganese content in hair samples collected from 228 Russian welders who performed CO2-shielded GMAW. The welders were divided into two groups. Persons in the first group welded low and medium alloy steel. They worked in open, well-ventilated areas of the factory. The second group used electrodes with a high manganese content and worked in poorly ventilated areas. Their manganese exposure was estimated to be seven to ten-times greater than that of workers in the first group.

Average manganese levels of less than 2 mg% and approximately 7.5 mg% were found in the hair of workers in the first and second groups, respectively. Early signs of manganese intoxication were seen only among workers in the second group. The investigators concluded that a direct relationship exists between the extent of manganese exposure and the concentration of manganese in hair. An upper allowable limit of manganese in the hair of 3-5 mg% was recommended. The amount of data provided by these investigators was very limited; neither statistical analyses nor concentration ranges were reported. Hence, it is difficult to judge the validity of the recommended limit.

Section Three Toxicologic Investigations in Animals and in Cell Cultures

23. Animal Studies

Many of the animal experiments performed during this report period investigated the early effects of welding particles in the lungs and the fibrogenic response elicited by welding fumes or their components.

Lam et al. (Ref. 79) studied the effects of inhaled zinc oxide fumes on the guinea pig lung. Animals were exposed for three hours per day for six days to 5 mg/m3 zinc oxide, a concentration equivalent to the TLV. Alterations in pulmonary function were noted during the first three days after exposure. These included a reversible increase in flow resistance, and reduction in lung compliance, carbon monoxide diffusing capacity and lung volumes. The lung weight was markedly increased and did not return to normal during the seventy two-hour post-exposure examination period.

Microscopic lesions observed in the lungs consisted of inflammation of the alveoli and interstitial thickening which was associated with infiltration of macrophages, lymphocytes, neutrophils and pulmonary fibroblasts. The elevation in lung weight may have been due to the cellular infiltrates in the lung, while the interstitial thickening and the presence of fibroblasts among the infiltrate may have caused the observed change in lung mechanics. The investigators concluded that pulmonary changes can occur with relatively few exposures to zinc oxide at the recommended TLV of 5 mg/m3. On the basis of this study, Lam et al. stated that this standard may not be low enough to protect exposed workers.

Sylvestre and P'an (Ref. 150) studied the effects of ozone, CO, NO2, and manganese dioxide (MnO2) on the mouse lung. Groups of mice were exposed by inhalation five hours a day for thirty days to (A) filtered air, (B) a mixture of ozone, CO, and NO2, (C) a mixture of ozone, CO, NO2, and MnO2, and (D) MnO2 alone. The concentrations were selected to simulate actual welding exposures. Animals were sacrificed and lungs examined at ten, twenty, and thirty days after the termination of the exposure.

All experimental groups had emphysema and dilated alveoli. Emphysematic lesions were less extensive in group C than in groups B and D. In animals exposed to MnO2, as well as to the mixture of ozone, CO, NO2, and MnO2, the bronchioles were filled with mucous. This effect was greatest in mice receiving MnO2 alone. The mixture of gases and particulates caused the most overall damage. Numerous areas of inflammation, edema, and dilated blood vessels were found only in the lungs of animals treated with the gases and MnO2. This study demonstrated that the combined effects of mixtures of gases and particles may differ from the effects of their individual components.

Gorban et al. (Ref. 50) compared the effects on the lungs of single exposures to particulates present in fumes from CO2-shielded GMAW of high alloy and low alloy steel. Nickel and chromium were present only in fumes from high alloy steel which also had three times more manganese than fumes from low alloy steel.

Fume samples were introduced into rat lungs by intratracheal instillation (a technique whereby materials are injected directly into the lung through a cannula which is passed into the pharynx and extends through most of the length of the trachea). Rats were sacrificed at one, three, and six months after treatment. Changes in the activity of serum enzymes and nucleic acid contents of the liver and lungs were observed. None of the welding particulates induced pulmonary fibrosis, as shown by the absence of differences in the hydroxyproline content (a unique constituent of collagen fibers) between the lungs of treated and control animals.

The LD_{50}'s, determined by intraperitoneal injection, indicated that the acute toxicity of particulates decreased as the welding current increased and that fumes from high alloy steel were substantially more toxic than those from low alloy steel (Table 7). This data is in apparent conflict with the study of Olah and Tolgyessy, (Ref. 116 and 120) which indicated

Table 7
Variation of Acute Toxicity (LD_{50}) of Welding Fumes with Welding Current

	Current	Electrode Diameter mm	Species	LD_{50} (mg/kg)
high alloy	200–250	1.4	mouse	950
	300–350	1.4	mouse	1450
	300–350	2.0	mouse	2017
	400–450	2.0	mouse	1917
	300–350	1.4	rat	1000
low alloy	300–350	2.0	mouse	>5000

Data from Gorban et al., Ref. 50

that chromium and nickel concentrations increase with increasing arc current. Gorban did not provide information about changes in the chemical composition of fumes with arc current.

Weller and Reichel (Ref. 163) investigated the fibrogenicity of welding fume particulates by injecting welding fume samples into the rat peritoneal cavity. Fumes were collected at a steel engineering company. The welding processes from which fumes were collected were not specified, but the samples contained 88.5 percent iron, 7.5 percent manganese, 3.2 percent silicon, and 0.9 percent aluminum. Rats were sacrificed and examined at three, six, and twelve months after treatment.

Small black granulomas were observed in the omentum at all three time periods. Dust deposits were seen in the spleen but not in the liver or lymph nodes. A large number of macrophages and connective tissue fibers were associated with dust deposits after three months. The fibrosis did not progress during the remainder of the study. Weller and Reichel concluded that the discrete fibrotic changes in the peritoneum parallel changes seen in the human lung, and that welding dusts cause a limited fibrotic response which does not progress into massive fibrosis.

In a similar experiment, Mal'ik et al. (Ref. 85) observed no signs of fibrosis in guinea pigs injected intraperitoneally with particulates generated by welding with basic and acid electrodes. Only pigment deposition without fibrotic reactions were seen.

Olah et al. (Ref. 118) examined the lungs of Wistar rats twelve weeks after intravenous injection of 50 mg welding fumes collected from SMAW with five different electrodes and argon-shielded GTAW of austentitic stainless steel. SMAW fumes produced a mild to moderate fibrogenic response in the lungs, but no such reaction followed treatment with GTAW fumes. Because intravenous injection is so far removed from the route by which humans are exposed to welding fumes, the relevance of this data is not clear.

Kalliomaki et al. (Ref. 69) compared pulmonary retention and clearance rates of nickel and chromium in fumes from GMAW and SMAW of stainless steel. Rats were exposed by inhalation to 40 mg/m3 fumes for one hour daily for four weeks. At the end of the exposure period, nickel and chromium generated by GMAW were present in higher concentrations in the lung than were nickel and chromium produced by SMAW (Figure 6). This difference reflects the chemical composition of the fumes to which the animals were exposed. Earlier reports by these investigators indicated that there is about four times as much nickel and about fifteen times as much chromium present in GMAW fumes than comparable quantities of SMAW fumes. Approximately the same ratios of nickel and chromium were found in lungs from animals treated with GMAW and SMAW fumes.

Clearance rates were investigated by periodically measuring concentrations of nickel and chromium in the lung for up to three months after the end of the four week exposure period. Nickel concentrations in lungs exposed to GMAW fumes were too low to yield reliable results. Nickel from GMAW fumes of stainless steel cleared rapidly at first with a half-life of three days. After several days, a second clearance pattern with a half-life of forty days was evident. Chromium from SMAW fumes cleared from the lungs with a half-life of forty days, but chromium from GMAW fumes cleared much more slowly, with a half-life of approximately 240 days. The investigators could not account for this difference on the basis of the available information on the solubility of chromium compounds in welding fumes. Nickel was excreted more rapidly than chromium. Chromium from GMAW fumes was eliminated in the urine in a bimodal pattern. At first, chromium levels fell rapidly in the urine, with a half-life of eight hours. Two days after exposure ceased, urinary chromium excretion decreased to a half-life of thirty days.

24. In Vitro Studies

Tests of the genetic effects of substances in bacterial or cell cultures offer a means for identifying compounds or mixtures of chemicals which may possibly be mutagenic to higher organisms. In addition, certain tests using cultured mammalian cells can detect oncogenic (tumor-producing) changes within the cells. The activity of genotoxic compounds is influenced by many factors within intact organisms, such as enzymatic detoxification or activation of foreign chemicals and destruction of genetically altered cells by the immune system. Therefore, in vitro assays are only capable of indicating whether or not a chemical is potentially genotoxic or carcinogenic to animals. Positive findings must be verified by animal studies before suspect compounds can be considered to be potential human carcinogens.

24.1 Bacterial Assays. The Salmonella/Ames test is widely used as a screening test for mutagenicity and as a pre-screening test for carcinogenicity studies. In the past, most such tests performed with welding

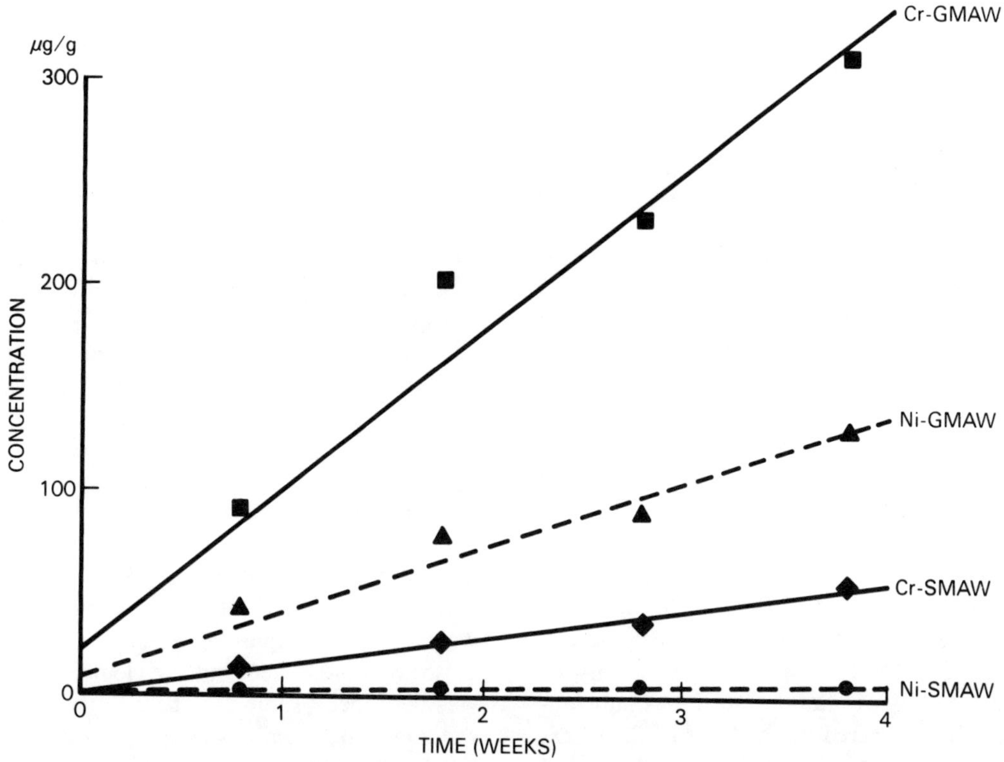

Redrawn from Kalliomaki et al., (Ref. 69)

Figure 6 — Retention of Nickel and Chromium in the Lungs of Rats Exposed to SMAW/SS and GMAW/SS Welding Fumes

fumes indicated that only stainless steel welding fumes are mutagenic. Hexavalent chromium has been implicated as the mutagenic component since particulates from stainless steel welding fumes containing 15–25 percent chromium were mutagenic while particulates of fumes from mild steel welding, containing less than 0.1 percent chromium, were not.

Unlike previous reports, Biggart (Ref. 19) found that both the gas phase and the particulates from mild steel welding fumes contained mutagens. The mutagenicity of fumes from mild steel welding was assessed with a modified Salmonella/Ames test, which purportedly offered increased sensitivity. Modifications of the test included preincubation of the bacteria with the test material, toxicity measurements, elimination of the top agar layer, and variations in cell density. Using these modifications, particulates from mild steel welding fumes were found to contain both direct-acting and indirect-acting mutagens (i.e., indirect-acting mutagens require activation by mammalian metabolic enzymes before mutagenicity is expressed.) The gas phase of mild steel welding fumes was potently mutagenic. According to Biggart, the mutagenicity was due only in small part to nitrogen dioxide in the fumes, but no further explanation was given. This data could potentially be quite important but, because only a short abstract with little detail and virtually no data was published, it is not possible to judge its significance.

24.2 Mammalian Cell Studies. Hansen and Stern (Refs. 57 and 105) studied the cytotoxicity and transforming effects of welding fumes and components of welding fumes on cultured baby hamster kidney cells (BHK-21) and Syrian hamster embryo (SHE) cells. Toxicity was measured in terms of the fraction of the cultured cells which survived exposure to test materials. Cell transformation is an indication of genetic and possibly oncogenic changes in the cell. It is detected by alterations in the cell morphology and growth characteristics of the cell culture. Untransformed cells usually grow as a confluent monolayer on the bottom of the culture flask, whereas transformed cells are more irregular and may pile up on top of each other forming small colonies. Such colonies may be oncogenic. This can be tested by injecting transformed cells into animals to determine whether they develop into malignant tumors. Animal tests were not performed in the studies described below.

When based on the nickel content, the degree of transformation of BHK-21 cells was the same for a variety of nickel compounds and for fumes from GMAW of nickel. Similar tests with chromium compounds and chromium containing GMAW fumes indicated that only hexavalent chromium compounds were cytotoxic or transforming; i.e., neither chromium metal nor trivalent chromium caused cell transformation. The transforming potency and toxicity of fumes from SMAW of stainless steel were ten times greater than those from GMAW of stainless steel. Fumes from SMAW and GMAW of mild steel were only very weakly toxic and did not cause cell transformation. In the SHE system, the toxicity of fumes from GMAW of stainless steel was substantially greater than expected on the basis of their hexavalent chromium content (Ref. 58). The increased toxicity was presumably due to the nickel content of the fumes. The toxicity and transforming effects of fumes from SMAW of stainless steel, which did not contain any detectable nickel, corresponded to levels expected from their soluble hexavalent chromium content.

As a follow-up to previous studies which indicated that fumes collected by impingement in water had higher concentrations of hexavalent chromium than did those collected with filters (Refs. 53 and 156), Hansen and Stern (Ref. 57) compared the transforming effects of fumes collected by impingement techniques and by filters. Fumes from GMAW of stainless steel collected by impingement in water contained five to ten times more hexavalent chromium than did fumes collected on filters. When toxicity was considered in terms of hexavalent chromium content, fumes collected in water were two to three times more toxic than fumes collected by impingement in cell culture medium.

The authors speculated that the difference in activity between samples collected by impingement in water and culture medium was due to the sequestering of the active compound(s) by ligand formation with organic compounds. In addition, Hansen and Stern (Ref. 57) found that fumes from GMAW of nickel collected in medium were ten times more soluble than fumes collected in water. However, the soluble nickel fraction is not toxic or transforming. This is in apparent conflict with earlier reports by Niebuhr, Stern et al. (Refs. 111 and 112) that the serum soluble fraction of fumes from GMAW of mild steel or cast iron using nickel rich electrodes caused more chromosomal damage per mol nickel in sister chromatid exchange assays than did the water-soluble fractions from the welding fumes. How these different responses to serum soluble nickel in the two assay systems have been resolved by Stern and his co-workers is not known.

Potebnia et al. (Ref. 127) tested the toxicity and co-carcinogenicity of welding fumes in SA7 adenovirus-infected cultured rat kidney cells and hamster embryo cells. This test system detects the enhancement of SA7 adenovirus-induced oncogenic transformation of cultured cells by the simultaneous exposure to certain co-carcinogens (substances that may not in themselves be carcinogenic but may enhance the activity of carcinogenic agents). Three welding fume samples were tested. The welding processes were not given, but the elemental composition was described (Table 8). The relative toxicity of the welding fume samples was sample number 2> number 1> number 3.

All three fume samples enhanced the oncogenic transformation of both cell lines. The optimum concentration for the enhancement of cell transformation was 4 ug/ml. There were no substantial differences in the co-carcinogenic potency among the welding fume samples. Transformation of hamster embryo cells was enhanced between 2.3 to 2.5 times by all the welding fumes. The transformation rates of

Table 8
Concentration of Elements in Three Welding Fume Samples Tested in the SA7-Adenovirus Cell Transforming System

	Concentration (Percent)		
Element	Sample 1	Sample 2	Sample 3
Manganese	30	5	4.4
Nickel	3	36	1.3
Iron	20	30	14.6
Copper	–	2.7	–
Trivalent chromium	5.5	0.7	7.5
Hexavalent chromium	0.03	–	–
Cr_2O_3			21.8
CrO_3			11.5

Data from Potebnia et al., Ref. 127

rat kidney cells were enhanced by 1.5, 2.1 and 2.6-fold by welding samples number 1, 2 and 3 respectively. No cell transformation occurred when cells were treated with welding fume samples in the absence of adenovirus. When transformed cells were injected into animals, tumors developed more rapidly in animals treated with welding fumes and virus than in those treated with virus alone. The design of the experiments did not enable conclusions to be drawn about the contributions of individual fume components to the co-carcinogenicity of the samples.

References

(1) ACGIH, Threshold limit values and biological exposure indices for 1986–1987. American Conference of Governmental Industrial Hygienists, Cincinnati, OH, p. 44, 1986

(2) Aitio, A. "Biological Monitoring of Occupational Exposure to Nickel." In: Sunderman, F.W., Jr., et al., (eds.), Nickel in the Human Environment, IARC Sci. Publ. No. 53., Lyon, France, pp. 497–505, 1984

(3) Akbarkhanzadeh, F. "A Personal Sampler for Measuring Shift Exposure to Welding Fumes." Ann. Occup. Hyg., 28(4):365–371 (1984)

(4) Akesson, B. and Skerfving, S. "Exposure in Welding of High Nickel Alloy." Int. Arch. Occup. Environ. Health, 56(2):111–117 (1985)

(5) American Welding Society "Laboratory Method for Measuring Fume Generation Rates and Total Fume Emission of Welding and Allied Processes." Miami, FL, American Welding Society, ANSI/AWS F1.2-85, 198

(6) American Welding Society "Method for Sampling Airborne Particulates Generated by Welding and Allied Processes." Miami, FL, American Welding Society, ANSI/AWS F1.1-85, 1985

(7) Anatovskaia, V.S. "Status of the Nervous System in Chronic Bronchitis." Vrach. Delo, (2):88–91 (1984)

(8) Anonymous "Welding and Cutting Tests for the Assessment of the Toxicity of Paint Primer Thermal Decomposition Products." Br. Weld. J., 15:468–469 (1968)

(9) Anonymous "Why Welding Fumes Needn't be a Problem?" Can. Mach. Metalwork (Canada), 80(1):12–14,35 (1984)

(10) Artem'ev, V.N. and Mol'kova, V.L. "Physiologic and Hygienic Assessment of Overalls Made of Various Fabrics and Materials." Gig. Tr. Prof. Zabol., (4):36–39 (1984)

(11) AV Svetskommissionen Ingenjorsvetenskapsakademien "Swedish Standard for Classification of Manual Metal Arc Electrodes into Fume Classes." 1975

(12) Baker, R.S.U., Arlauskas, A., Tandon, R.K., Crisp, P.T. and Ellis, J. "Toxic and Genotoxic Action of Electric-Arc Welding Fumes on Cultured Mammalian Cells." J. Appl. Toxicol., 6:357-362 (1986)

(13) Bariffi, F., Marangio, E., Pesci, A., Bertorelli, G. and Gabrielli, M. "Bronchoalveolar Lavage in the Diagnosis of Inorganic Dust Induced Pulmonary Diseases." Riv. Patol. Clin. Tuberc. Pneumol., 56(3):337–345 (1985)

(14) Barnhart, S. and Rosenstock, L. "Cadmium Chemical Pneumonitis." Chest, 86(5):789–791 (1984)

(15) Barnhart, S., Rosenstock, L., Heyer, N., Pierson, D. and Hudson, L. "Assessment of Respiratory Sysptoms and Bronchodilator Responsiveness among Welders." Chest, 86(2):302 (1984)

(16) Bartley, D.L., McKinnery, W.N., Jr. and Wiegand, K.R. "Ultraviolet Emissions from the Arc-Welding of Aluminum-Magnesium Alloys." Am. Ind. Hyg. Assoc. J., 42(1):23–31 (1981)

(17) Baselt, R.C. "Biological Monitoring Methods for Industrial Chemicals." Biomedical Publications, Davis, California, 1980

(18) Becker, N., Claude, J. and Frentzel-Beyme, R. "Cancer Risk of Arc Welders to Fumes Containing Chromium and Nickel." Scand. J. Work Environ. Health, 11(2):75–82 (1985)

(19) Biggart, N.W. "The Mutagenicity of Gases Particulates in Arc Welding Fumes." Diss. Abstr. Int. B 45(9):2884 (1985); 154 pp. (1984)

(20) Bjorseth, O., Jakobsen, K. and Syversen, T. "Hyperbaric Occupational Hygiene and Toxicology." Govt. Reports Announcements & Index (GRA&I), Issue 04; U.S. NTIS/PB85-121572, 31 pp. (1985)

(21) Bjorseth, O., Jakobsen, K. and Syversen, T. "Hyperbaric Occupational Hygiene and Toxicology." Foundation of Scientific and Industrial Research, Norwegian Institute of Technology, 1984

(22) Blakely, J.H. and Zatka, V.J. "The Determination of Hexavalent Chromium, Total Chromium, and Total Nickel in Welding Fume." J.R. Gordon Research Laboratory Report, INCO Metal Co., Mississauga, Ontario, Canada, 1978

(23) Blumthaler, M., Ambach, W. and Daxecker, F. "Calculation of Radiation Dose in Changes in Solar UV-B Radiation with Regard to Keratitis Photoelectrica." Klin. Monatsbl. Augenheilkd., 186(4):275–278 (1985)

(24) Chiesura, P., Trevisan, A., Gori, G.P., Buzza, A. and Calzavara, V. "On the Risk of Cadmium Poisoning in Brazing Operations." Med. Del Lav., 75(4):300–305 (1984)

(25) Christensson, G. and Rockert, H. "Effects of

Electrical Welding under Water on Dental Amalgams and Enamel In Vitro." Int. Res. Comm. System, 8:756-757 (1980)

(26) Clayton, G.D. and Clayton, F.E. (eds.) "Patty's Industrial Hygiene and Toxicology." Third Revised Edition, Volume 2, John Wiley and Sons, Inc., New York, 1981

(27) Coenen, W., Grothe, I., Kuehnen, G., Pfeiffer, W., Schenk, H. and Augustin, S. "Welding Fume Exposure at Work Places. Nickel and Chromate in Welding Fumes—Part 2." Staub—Reinhalt. Luft, 45(12):546-550 (1985)

(28) Dare, P.R.M., Hewitt, P.J., Hicks, R., Van Bemst, A., Zober, A. and Fleischer, M. "Barium in Welding Fumes." Ann.

(29) Dare, P.R.M., Hewitt, P.J., Hicks, R., et al. "Analysis of Chromium in Welding Fumes." Ann. Occup. Hyg., 28(2):275-277 (1984)

(30) Diebold, F., Salsi, S., Aubertin, G., Cicolella, A. and Limasset, J.C. "Exposure to Welding Gases and Dusts—Measurement of Air Contamination in Two French Factories. Technical Prevention." Cahiers de Notes Documentaires—S'ecurit'e et Hygi'ene du Travail 3rd Quarter, No. 120, Note No. 1536-120-85, pp. 307-316 (1985)

(31) Dobromyslov, A.N., Fedorova, I.D. and Evdokimova, Iu. A. "Evaluation of the Functional Capacity of the Macular Area According to the Results of the Photo-Stress Test in Welders." Vestn. Oftalmol., 101(5):33-36 (1985)

(32) Doorgeest, T. "Antifouling Paints" Vejligheid, 60(12):629-638 (1984)

(33) Doorgeest, T. "Paint Products Containing Isocyanates." Vejligheid, 60(8):391-399 (1984)

(34) Dvorak, J., Svabova, K. and Malek, B. "Effect of the Electric Arc on the Front of the Eye during Welding." Pracovni Lekarstvi, 36(4):137-138 (1984)

(35) Elinder, C.G., Edling, C., Lindberg, E., Kagedal, B. and Vesterberg, O. "Beta-2-microglobulinuria among Workers Previously Exposed to Cadmium: Follow-Up and Dose-Response Analyses." Am. J. Ind. Med., 8(6):553-564 (1985)

(36) Elinder, C.G., Edling, C., Lindberg, E., Kagedal, B.K. and Vesterberg, O. "Assessment of Renal Function in Workers Previously Exposed to Cadmium." Br. J. Ind. Med., 42(11):754-760 (1985)

(37) Eriksen, P. "Time Resolved Optical Spectra from MIG Welding Arc Ignitions." Am. Ind. Hyg. Assoc. J., 46(3):101-104 (1985)

(38) Esnault, A., Henry, D., Riet, D. and Yhuel, C. "Mortality in a Group of a Hundred Welders, Exposed During 30 Years to Welding Fumes." Arch. Mal. Prof. Med. Trav. Secur. Soc. (France), 46(6):440-441 (1985)

(39) Eun, H.C., Kim, K.C. and Cha, C.W. "Occupational Burns." Contact Dermatitis, 10:20-22 (1984)

(40) Falck, F.Y., Keren, D.F., Fine, L.J., Smith, R.G. and McClatchey, K.D. "Protein Excretion Patterns in Cadmium-Exposed Individuals: High Resolution Electrophoresis." Arch. Environ. Health, 39(2):69-73 (1984)

(41) Farkas, I. "WHO-Coordinated International Study on the Health Effects of Occupational Exposure of Welders to Chromium and Nickel." Nutr. Res., Suppl. I, pp. 683-689 (1985)

(42) Felzmann, E. "Effect of Surface Metallization of Goggles on Their Eye-Protective Properties under the Action of Heat Radiation." ZIS Mitt., 27(6):626-634 (1985)

(43) Fog, H.M. and Rietz, B. "Piezoelectric Crystal Detector for the Monitoring of Ozone in Working Environments." Anal. Chem., 57(13):2634-2638 (1985)

(44) Frentzel-Beyme, R. "Epidemiology in Occupational Medicine in South East Asia." Int. Arch. Occup. Environ. Health, 54(4):365-368 (1984)

(45) Gerin, M., Siemiatycki, J., Richardson, L., Pellerin, J., Lakhani, R. and Dewar, R. "Nickel and Cancer Associations from a Multicancer Occupation Exposure Case-Referent Study: Preliminary Findings." In: Sunderman, F.W., Jr., et al., (eds.), Nickel in the Human Environment, IARC Sci. Publ. No. 53., Lyon, France, pp. 105-115, 1984

(46) Glinsmann, P.W. and Rosenthal, F.S. "Evaluation of an Aerosol Photometer for monitoring Welding Fume Levels in a Shipyard." Am. Ind. Hyg. Assoc. J., 46(7):391-395 (1985)

(47) Goller, J.W. and Paik, N.W. "A Comparison of Iron Oxide Fume inside and outside of Welding Helmets." Am. Ind. Hyg. Assoc. J., 46(2):89-93 (1985)

(48) Golovatjuk, A.P., Levcenko, O.G., Veblaja, T.S., Zykov, G.A., Harcanko, N.P. and Sevcenko, L.A. "Use of Physical Methods for Studying the Composition of Solid Constituents of Welding Fumes." Svarocnoe Proizvodstvo, (3):37-38 (1984)

(49) Gorban, L.N., Krasnyuk, E.P. and Lubyanova, I.P. "Possibility of Using Data on the Content of Manganese in the Hair of Welders as an Exposure Test." Vrach. Delo, (6):87-89 (1985)

(50) Gorban, L.N., Prilipko, V.A. and Cherednichenko, V.M. "Data on the Comparative Toxicity of Aerosols Formed in the Welding of High-Alloy and

Low-Alloy Steels." Gig. Tr. Prof. Zabol., (5):16–19 (1984)

(51) Gos, R., Horowski, P. and Stepien, J. "Functional Examinations of the Sight Organ of Welders of the M-5 Division Brown Coal Mine Belchatow Poland." Med. Pr., 35(3):223–226 (1984)

(52) Gos, R., Stepien, J. and Horowski, P. "Evaluation of the Eyes of Welders at the M-5 Division Brown Coal Mine Belchatow Poland." Med. Pr., 35(2):133–136 (1984)

(53) Gray, C.N., Goldstone, A.M., Dare, P.R.M. and Hewitt, P.J. "The Evolution of Hexavalent Chromium in Metallic Aerosols." Am. Ind. Hyg. Assoc., 44:384–388 (1983)

(54) Grothe, I., Kuhnen, G. and Pfeiffer, W. "Workplace Measurements of Welding Fumes with Special Consideration of Nickel and Chromates" In: Welding and Cutting c84, Deutscher Verlag fur Schweisstechnik GmbH, Dusseldorf, Federal Republic of Germany, pp. 65–69, 1984

(55) Gullvag, B., Mylius, E. and Nilsen, A. "Counting Alveolar Macrophages (AM) from Expectorate Samples of Exposed Workers as a Test for Lung Irritation from Occupation Exposure." Bull. Environ. Contam. Toxicol., 34(5):702–708 (1985)

(56) Hallne, U. "Working Environment Problems Connected with Welding—Part 24: Measurement of Gaseous Air Contaminants during Gas-Shielded Arc Welding of Parts Degreased with Methylchloroform." Arbetarskyddsstyrelsen, Publikationsservice, 171 84, Solna, Sweden, 26 pp., 1984

(57) Hansen, K. and Stern, R.M. "Nickel and Chromium Compounds and Welding Fumes in Mammalian Cell Transformation Bioassay In Vitro." In: Stern., R.M., Berlin, A., Fletcher, A. and Jarvisalo, I., (eds.), Biological Effects and Health Hazards of Welding Fumes and Gases, Elsevier Biomedical Press, Amsterdam, pp. 305–310, 1986

(58) Hansen, K. and Stern, R.M. "Welding Fumes and Chromium Compounds in Cell Transformation Assays." J. Appl. Toxicol., 5(5):306–314 (1985)

(59) Herberts, P., Kadefors, R., Andersson, G. and Petersen, I. "Shoulder Pain in Industry: An Epidemiological Study on Welders." Acta Orthop. Scand., 52(3):299–306 (1981)

(60) Herberts, P., Kadefors, R., H'ogfors, C., and Sighoim, G. "Shoulder Pain and Heavy Manual Labor" Clin. Orthop., (191):166–178 (1984)

(61) Hernberg, S. "Lead." In: Aitio, A., Raiihimake, V., and Vainio, H., Biological Monitoring and Surveillance of Workers Exposed to Chemicals, Hemisphere Publishing Corp., pp. 19–27, 1984

(62) Huel, G., Everson, R.B. and Menger, I. "Increased Hair Cadmium in Newborns of Women Occupationally Exposed to Heavy Metals." Environ. Res., 35:115–121 (1984)

(63) Il'nitskaya, A.V. and Kalina, O.V. "Health Effects of Dust-Gas Mixtures in Plasma Technology." Gig. Tr. Prof. Zabol., (10):5–9 (1980)

(64) International Labour Office "Encyclopaedia of Occupational Health and Safety." International Labour Office, Geneva, Two Volumes, Third (Revised) Edition, Technical Editor: Dr. Luigi Parmeggiani, 1983

(65) Iwamoto, N., Umesaki, N., Kamai, M., Kobayshi, M., Tsutsumi, S., Taga, M. and Kume, A. "Behavior of Manganese in Welding Fume (I)." Trans. JWRI, 13(1):21–26 (1984)

(66) Iwao, S., Baba, Y. and Kodama, Y. "Disappearing Workers in Dust Operations" Scand. J. Work Environ. Health, 10:138 (1984)

(67) Jenkins, N. and Moreton, J. "Letters to the Editor." Am. Ind. Hyg. Assoc. J., 45(9):B-14 (1985)

(68) Johnston, A.M., Vincent, J.H. and Jones, A.D. "Measurements of Electric Charge for Workplace Aerosols." Ann. Occup. Hyg., 29(2):271–284 (1985)

(69) Kalliomaki, P.-L., Olkinuora, M., Hyvarinen, H.-K. and Kalliomaki, K. " Kinetic of Nickel and Chromium in Rats Exposed to Different Stainless-Steel Welding Fumes." In: Sunderman, F.W., Jr., et al., (eds.), Nickel in the Human Environment, IARC Sci. Publ. No. 53., Lyon, France, pp. 385–394, 1984

(70) Kalnas, J. and Alleyne, B.C. "A Health Assessment of Workers Exposed to Lead in Radiator Shops in Alberta—1983." Alberta Workers Health, Safety and Compensation, Occup. Health and Safety Division, 10709, Jasper Avenue, Edmonton, Alberta, T5J 3N3, Canada, 68 pp., 1984

(71) Karai, I. and Horiguchi, S. "Pterygium in Welders." Br. J. Ophthamol., 68(5):347–349 (1984)

(72) Karai, I., Matsumura, S., Takise, S., Horiguchi, S. and Matsuda, M. "Morphological Change in the Corneal Endothelium due to UV Radiation in Welders." Br. J. Ophthalmol., 68(8):544–548 (1984)

(73) Knight, G.S., Williams, H.E. and Hinton, D. "Elevated Plasma Manganese Levels in Welders Cutting Manganese Steel." N.Z. Med. J., 98(788):870 (1985)

(74) Knudsen, J. and Bjerre, A. "A Method of Hazard Assessment of a Gaseous Substance with Respect to Formation of Toxic Photodecomposition Products: Application to Carbon Tetrachloride, Trichlorofluoromethane and Dichlorodifluoromethane." Chemosphere, 14(2):249–255 (1985)

(75) Koshi, K. and Yagami, T. "Chromosomes of

Cultured Peripheral Lymphocytes from Stainless Steel Welders (Part 2)." Mutat. Res., 130:369 (1984)

(76) Koshi, K., Yagami, T. and Nakanishi, Y. "Hypogenetic Analysis of Peripheral Blood Lymphocytes from Stainless Steel Welders." Ind. Health (Japan), 22(4):305–318 (1984)

(77) Kovacs, I. "Chemical Air Pollution of Uncovered Metal Industry Operations and Reduction of Its Rats." Munkavedelem, Munka-Uzemegeszsegugy, 30(10–12):149–156 (1984)

(78) Lahti, T., Pekkarinen, J. and Starck, J. "New Noise Level Definition: Acting on Impulse." Work Health Safety, Institute of Occupational Health, Haartmaninkatu 1, 00290 Helsinki, Finland, pp. 26–27, 1984

(79) Lam, H.F., Conner, M.W., Rogers, A.E., Fitzgerald, S. and Amdur, M.O. "Functional and Morphologic Changes in the Lungs of Guinea Pigs Exposed to Freshly Generated Ultrafine Zinc Oxide." Toxicol. Appl. Pharmacol., 78(1):29–38 (1985)

(80) Langard, S. and Stern, R.M. "Nickel in Welding Fumes A Cancer Hazard to Welders? A Review of Epidemiological Studies on Cancer in Welders." In: Sunderman, F.W., Jr., et al., (eds.), Nickel in the Human Environment, IARC Sci. Publ. No. 53., Lyon, France, pp. 95–103, 1984

(81) Limpel, I., Hrzenjak, M. and Gradisar, B. "Amount of Fluorine in Fumes from Welding with Alkaline Coated Electrodes." Varjlna Teh., 34(2):33–34 (1985)

(82) Louhevaara, V., Tuomi, T., Smolander, O., Korhonen, A., Tossavainen, A. and Jaakkola, J. "Cardiorespiratory Strain in Jobs That Require Respiratory Protection." Int. Arch. Occup. Environ. Health, 55(3):195–206 (1985)

(83) Lundberg, I., Sjoegren, B., Hallne, U., Hedstroem, L. and Holgersson, M. "Environmental Factors and Uptake of Cadmium among Brazers Using Cadmium-Containing Hard Solders." Am. Ind. Hyg. Assoc. J., 45(6):353–359 (1984)

(84) Lydahl, E. and Philipson, B. "Infrared Radiation and Cataract. I. Epidemiologic Investigation of Iron-and Steel-Workers." Acta Ophthalmol. (Copenhagen), 62(6):961–975 (1984)

(85) Mal'ik, E., Ulrich, L., Humpl'ikov'a, J., Hurb'ankov'a, L., Chovan, L., Turzova, M. and Dornetzhuber, J. "Pneumoconiotic Changes in Welders." Studia Pneumologica et Phtiseologica Cechoslovaca, 44(5):307–316 (1984)

(86) Manasova, E. "Austenitic Electrode with Reduced Production of Fumes and Cr6+." Zvaranie, 33(7):198–201 (1984)

(87) Mantysalo, S. and Vuori, J. "Effects of Impulse Noise and Continuous Steady State Noise on Hearing." Br. J. Ind. Med., 41(1):122–132 (1984)

(88) Mao, Z., Zhang, H., Huang, C. and Zou, C. "Observation on Development of Pneumoconiosis among Manual Arc Welders in Dalian Shipyard (1960–1982)." Institute of Health, China National Center for Preventive Medicine, Beijing, China, 8 pp., 1984

(89) Marek, K. "Lung Function in Different Types of Pneumoconiosis." Z. Erkr. Atmungsorgane, 164(4):270–272 (1985)

(90) Mastromatteo, E. "Cancer in Relation to Occupational Exposure to Nickel." Carcinog Mutagens Environ., 5:3–15 (1985)

(91) McMillan, G. "Cleaner Metal—Cleaner Air." Occup. Health (London), 37(12):552–560 (1985)

(92) Milham, S., Jr. "Mortality in Workers Exposed to Electromagnetic Fields." Environ. Health Perspect., 62:297–300 (1985)

(93) Millodot, M. and Earlam, R.A. "Sensitivity of the Cornea after Exposure to Ultraviolet Light." Ophthalmic Res. (Switzerland), 16(6):325–328 (1984)

(94) Moreton, J. "Fume Emission Rates for Open Arc Welding Processes." Metal Construction, 17(5):308–310 (1985)

(95) Moreton, J., Bettelley, J., Mathers, H., Nicholls, A., Perry, R.W., Ratcliffe, D.B. and Svensson, L. "Investigations of Techniques for the Analysis of Hexavalent Chromium, Total Chromium, and Total Nickel in Welding Fume: A Cooperative Study." Ann. Occup. Hyg., 27:137–156 (1983)

(96) Moreton, J., Bettelley, J., Mathers, H., Svensson, L., Nicholls, A., Ratcliffe, D.B. and Perry, R.W. "Analysis of Chromium in Welding Fumes." Ann. Occup. Hyg., 28:276–277 (1984)

(97) Moreton, J. and Jenkins, N. "Pollutants from Paints when Welding Primed Steel—New Test Method Proposed. " Metal Construction, 16(10):605–607 (1984)

(98) Moreton, J. and Jenkins, N. "Barium in Welding Fumes (Letters to the Editor). " Ann. Occup. Hyg., 29:443–444 (1985)

(99) Moreton, J., Smars, E.A. and Spiller, K.R. "Fume Emission when Welding Stainless Steel." Metal Construction, 17(12):794–798 (1985)

(100) Moreton, J. and Stevens, D.C. "FED—A database to Assist with Welding Fume Problem." Weld. Inst. Res. Bull., 25(9):301–304 (1984)

(101) Morgenroth, K. and Verhagen-Schroeter, G. "Light Microscopic and Electron Microscopic Examination and Energy Dispersive Radiologic

Microanalysis of Biopsy Probes for the Pathogenesis of Arc-Welders Lung." Atemwegs-Lungenkrankh. (Germany, West), 10(9):451–456 (1984)

(102) Morris, B.W., Hardisty, C.A., McCann, J.F., Kemp, G.J. and May, T.W. "Evidence of Chromium Toxicity in a Group of Stainless Steel Welders." Atmos. Spectrosc., 6(6):149–151 (1985)

(103) Moseley, H. "Ultraviolet and Visible Radiation Transmission Properties of Some Types of Protective Eyewear." Phys. Med. Biol. (England), 30(2):177–181 (1985)

(104) Mueller, E.J. and Seger, D.L. "Metal Fume Fever—A Review." J. Emerg. Med., 2(4):271–274 (1985)

(105) Mur, J.M., Pham, Q.T., Meyer-Bisch, C., Massin, N. and Moulin, J.J. "Exposure To Welding Gases and Dusts—Results of Epidemiological Surveys." Cahiers de Notes Documentaires—S'ecurit'e et Hygi'ene du Travail 2nd Quarter, No. 119, Note No. 1530–119–85, pp. 233–238 (1985)

(106) Mur, J.M., Teculescu, D., Pham, Q.T., Gaertner, M., Massin, N., Meyer-Bisch, C., Moulin, J.J., Diebold, F., Pierre, F., et al. "Lung Function and Clinical Findings in a Cross-Sectional Study of Arc Welders: An Epidemiological Study." Int. Arch. Occup. Environ. Health, 57(1):1–18 (1985)

(107) Naumenko, I.M., Nemoshkalenko, V.V. and Piliankevich, A.N. "New Data on Physico-Chemical Properties of Welding Aerosols." Gig. Tr. Prof. Zabol., (12):5–7 (1984)

(108) Nemcova, A. "Hygiene of Cutting and Resistance Seam Welding Technologies." Zyaranie, 33(7):202–205 (1984)

(109) Nemcova, A. "Toxicity of the Gas Phase in Welding Fumes." Zvaracske Spravy, 35(1):10–17 (1985)

(110) Newhouse, M.L., Oakes, D. and Wooley, A.J. "Mortality of Welders and Other Craftsmen at a Shipyard in Northeast England (UK)." Br. J. Ind. Med., 42(6):406–410 (1985)

(111) Niebuhr, E., Stern, R.M., Thomsen, E. and Wulf, H.C. "Genotoxicity of Nickel Welding Fumes Assayed with Sister Chromatid Exchange in Human Blood Cells in Culture." Ann. Clin. Lab. Sci., 11(1):85–86 (1981)

(112) Niebuhr, E., Stern, R.M., Thomsen, E. and Wulf, H.C. "Relative Solubility of Nickel Welding Fume Fractions and Their Genotoxicity in Sister Chromatid Exchange In Vitro." U.S. NTIS, PB82–158411, 13 pp., c1980

(113) NIOSH "Criteria for a Recommended Standard . . . . Occupational Exposure to Inorganic Nickel." National Institute for Occupational Safety and Health, Washington, D.C., (DHEW (NIOSH) Publ. No. 77–164), 282 pp., 1977

(114) Nock, T. "Welding Fumes and Gases." Plant Eng. (England), 29(5):21–22 (1985)

(115) Okuno, T. "Spectra of Optical Radiation from Welding Arcs." Ind. Health, 23(1):53–70 (1985)

(116) Olah, L. "Heavy Metals in Welding Fumes." Jad. Energ., 30(10):364–365 (1984)

(117) Olah, L. "Radioanalytical Determination of Chromium in Welding Fumes." Jad. Energ., 31(2):47–49 (1985)

(118) Olah, L. and Kosnac, L. "Welding Aerosols and Their Physical and Biological Characteristics." Zyaranie, 33(7):205–209 (1984)

(119) Olah, L. and Pivoluska, J. "Determination of Chromium (VI) in Welding Fumes by Radionuclide X-Ray Fluorescent Analysis." J. Radioanal. Nucl. Chem. (Switzerland), 96(6):575–584 (1985)

(120) Olah, L. and Tolgyessy, J. "Determination of Heavy Metals in Welder's Working Environment Using Nuclear Analytical Methods." J. Radioanal. Nucl. Chem., 93(1):43–53 (1985)

(121) Olander, L. "Welding Fume Buoyant Plume." Aerosol. Sci. Technol., 4(3):351–358 (1985)

(122) Ortendahl, T.W., Dahlen, G. and Rockert, H.O.E. "Evaluation of Oral Problems in Divers Performing Electrical Welding and Cutting under Water." Undersea Biomed. Res., 12(1):69–76 (1985)

(123) Palmer, W.G. "Effects of Welding on Health IV." Miami, FL, American Welding Society, 1983

(124) Pantucek, M.B. "Hygienic Assessment of Fumes Formed during Welding of Galvanized Steel Sheets and during Galvanizing." Koroze Ochr. Mater., 28(3):62–65 (1984)

(125) Pantucek, M.B. "Hygienic Evaluation of Fumes Formed in Spot Welding of Galvanized Steel Sheets." Zyaranie, 33(7):220–222 (1984)

(126) Piljankevic, A.N. and Naumenko, I.M. "Electron Microscocy of the Soil Phase of the Aerosol from Electric Welding." Svarocnoe Proizvodstvo, (3):39 (1984)

(127) Potebnia, G.P., Struk, V.I. and Gorban, L.N. "Increase in the Transforming Activity of Adenovirus SA7 by Metal Aerosols in Cultures of Sensitive Cells." Vrach. Delo, (1):112–114 (1984)

(128) Preusz, H.J. and Geyer, O.C. "Contact Lenses. A Risk While Arc Welding?" Contactologia (Germany, West), 6(2):56–59 (1984)

(129) Protor, N.H. and Hughes, J.P. "Chemical Hazards of the Workplace." J.B. Lippincott Company, Philadelphia, PA (1978)

(130) Reichel, G. "Epidemiological Investigation Regarding the Frequency of Bronchial and Lung Changes in Arc Welders." Atemwegs-Lungenkrankh. (Germany, West), 10(9):444–447 (1984)

(131) Reichel, G. "New Aspects Regarding the Problem of Arc Welders' Lung." Atemwegs-Lungenkrankh. (Germany, West), 10(9):442–443 (1984)

(132) Reichel, G. "Pneumoconiosis by Inorganic Dusts." Atemwegs-Lungenkrankh. (Germany, West), 11(6):253–259 (1985)

(133) Richter, K.H. "On the Industrial Hygienic Situation of Arc Welders in the Chemical Industry." Z. Gesamte Hyg. Ihre Grenzgeb. (Germany, East), 31(11):628–631 (1985)

(134) Richter, K.H. and Ruess, U. "Initial Results of Radiant Heat Exposure of Arc Welders and Gray Cast Welders During Work." Z. Gesamte Hyg. Ihre Grenzgeb. (Germany, East), 31(9):536–538 (1985)

(135) Ross, D.S. "Dichloroacetyl Chloride Exposure? Case History." Occup. Health (London), 37(3):116–118 (1985)

(136) Schellhas, H. "Medical Investigations into the Hazards of Chromium and Nickel Containing Dusts in the Further Processing of Alloy Steels." Stahl Eisen, 14(84):43–45 (1984)

(137) Schneider, W.D., Grund, W., Liebich, R., Maintz, G. and Schmidt, G. "Epidemiological Cross-Sectional Studies of Occupational Pathology of Welders." Gigiena Truda (Kiev), (20):8–12 (1984)

(138) Silverstein, M., Maizlish, N., Park, R. and Mirer, F. "Mortality among Workers Exposed to Coal Tar Pitch Volatiles and Welding Emissions: An Exercise in Epidemiologic Triage." Am. J. Publ. Health, 75(11):1283–1287 (1985)

(139) Sjogren, B. and Ulfvarson, U. "Respiratory Symptoms and Pulmonary Function among Welders Working with Aluminum, Stainless Steel and Railroad Tracks." Scand. J. Work Environ. Health, 11(1):27–32 (1985)

(140) Sjogren, B., Lidums, V., Haekansson, M. and Hedstroem, L. "Exposure and Urinary Excretion of Aluminium During Welding." Scand. J. Work Environ. Health, 11(1):39–43 (1985)

(141) Smirnov, K.M., Viru, A.A. and Smirnova, T.A. "Comparative Analysis of Urinary Corticosteroid and Catecholamine Levels in Subjects with Different Professional Preparation." Fiziol. Chel., 11(5):834–838 (1985)

(142) Spacilova, M. and Hykes, P. "Pneumoconiosis due to Welding." Studia Pneumologica et Phtiseologica Cechoslovaca, 44(7):474–478 (1984)

(143) SRI International "Effects of Welding on Health V." Miami, FL, American Welding Society, (In Press)

(144) Stanulla, H. and Liebetrau, G. "Electric Welder's Lung." Prax. Klin. Pneumol., 38(1):14–18 (1984)

(145) Stern, R.M., Berlin, A., Fletcher, A., Hemminki, K., Jarvisalo, J. and Peto, J. "International Conference on Health Hazards and Biological Effects of Welding Fumes and Gases, Copenhagen, 18–21 February 1985, Summary Report." Int. Arch. Environ. Health, 57:237–246 (1986)

(146) Stern, R.M., Drenck, K., Lyngenbo, O., Dirksen, H. and Groth, S. "Thoracic Magnetic Moment, Occupational Exposure, and Health of Shipyard Welders." The Danish Welding Institute (Svejsecentralen), Report No. 85.61, Publ. No. Dk-2605, Brondby, Denmark, 40 pp., 1985

(147) Stokinger, H.E. "A Review of World Literature Finds Iron Oxides Non Carcinogenic." Am. Ind. Hyg. Assoc. J., 45(2):127–133 (1984)

(148) Stone, W.E. "Occupational Repititive Strain Injuries." Aust. Fam. Physician, 13(9):681–684 (1984)

(149) Suzuki, Y. and Serita, F. "Simultaneous Determination of Water-Soluble Trivalent and Hexavalent Chromium by Anion Exchange High-Pressure Liquid Chromatography." Ind. Health, 23(3):207–220 (1985)

(150) Sylvestre, D. and P'an, A. "Biological Effects of Some Pollutants in Welding Fumes. Effects of a Mixture of Zone, Nitrogen oxide (NO2), and Carbon Monoxide Plus Manganese Oxide (Mn02) Particles on the lungs of Mice." Union Med. Can., 113(1):9–10, 12–15 (1984)

(151) Tandon, R.K., Crisp, P.T., Ellis, J. and Baker, R.S. "Investigation of Hardfacing and HSLA Steel Electrodes." Weld. Met. Fabr., 53(2):43–47 (1985)

(152) Tandon, R.K., Crisp, P.T., Ellis, J., Baker, R.S. and Chenhall, B.E. "Chemical Investigation of Some Electric Arc Welding Fumes and Their Potential Health Effects." Aust. Weld. Res., 13:55–60 (1984)

(153) Tandon, R.K., Payling, R., Chenhall, B.E., Crisp, P.T., Ellis, J. and Baker, R.S. "Application of X-Ray Photoelectron Spectroscopy to the Analysis of Stainless-Steel Welding Aerosols." Appl. Surf. Sci., 20(4):527–537 (1985)

(154) Tanninen, V.P., Hyvarinen, H.K., Grekula, A. and Kalliomaki, P.L. "Experimental Improvement in Analysis of Aerosol Samples by X-Ray Powder Diffraction." J. Aerosol. Sci., 16(5):373–378 (1985)

(155) Tanninen, V.P., Mikkola, E., Hyvarinen, H.K., Grekula, A. and Kalliomaki, P.L. "Determination of the Valence of the Iron in Welding Fumes with the Kbeta Emission Spectrum." X-Ray Spectrom., 14(4):188–191 (1985)

(156) Thomsen, E. and Stern, R.M. "Collection, Analysis and Composition of Welding Fumes." Danish Welding Institute Report 81–09, IWW Doc. VIII-954–81, 1981

(157) Thorne, B.D. and Hewitt, P.J. "Generation of Cadmium Fume from Alloy Surfaces." Ann. Occup. Hyg. 2nd Quarter, 29(2):181–189 (1985)

(158) Tigges, K.D. "Measurements of Gases Generated during Arc Welding with Coated Electrodes." Staub—Reinhalt. Luft, 44(7–8):332–336 (1984)

(159) Trevisan, A. and Bonadonna, A. "On a Case of Cadmium Nephropathy." Med. Del Lav., 75(4):322–327 (1984)

(160) Van der Wal, J.F. "Exposure of Welders to Fumes, Chromium, Nickel, Copper and Gases in Dutch Industries." Ann. Occup. Hyg., 29(3):377–389 (1985)

(161) Villaume, J.E., Wasti, K., Liss-Suter, D. and Hsiao, S. "Effects of Welding on Health, Volume I." Miami, FL, American Welding Society, 1979

(162) Waksvik, H. and Boysen, M. "Cytogenic Analyses of Lymphocytes from Workers in a Nickel Refinery." Mutat. Res., 103:185–190 (1982)

(163) Weller, W. and Reichel, G. "The Effect of Dust Occurrence during Arc Welding Procedures: Intraperitoneal Test with Rats." Atemwegs-Lungenkrankh. (Germany, West), 10(9):448–450 (1984)

(164) Williams, M.K. "Biological Tests of Lead Absorption Following A Brief Massive Exposure." J. Occup. Med., 26(7):532–533 (1984)

(165) You, A., Li, C. and Yan, Y. "Gaseous Fluoride in Welding Fumes of Alkaline Electrodes." Hanjie Xuebao, 5(4):171–178 (1984)

(166) Zatka, V.J. "Speciation of Hexavalent Chromium in Welding Fumes Interference by Air Oxidation of Chromium." Am. Ind. Hyg. Assoc. J., 46(6):327–331 (1985)

(167) Zober, A. "Arbeitsmedizinische Untersuchungen zur Inhalativen Belastung Von Lichtbogenschmelzschweibern." Verlag fur Neue Wissenschaft, Bremerhaven, 1982

(168) Zober, A. "Study of the Kinetics of Chromium and Nickel in Biological Material during A Week of Arc Welding Using Chromium-Nickel-Containing Filler Metals." Schweissen und Schneiden, (10):1–3 (1984)

(169) Zober, A. "Symptoms and Findings at the Broncho Pulmonary System of Electric Arc Welders. 1. Epidemiology." Zentralbl. Bakteriol., 173(1–2):92–119 (1981)

(170) Zober, A. "The Variability of Biological Parameters and Lung Function Data during One Week of Welding Activity with Cr-Ni-Containing Weld Materials." In: Welding and Cutting c84, Deutscher Verlag fur Schweisstechnik GmbH, Dusseldorf, Federal Republic of Germany, pp. 69–72, 1984

(171) Zober, A., Schaller, K.H. and Weltle, D. "Inhalative Load and Stress of TIG-Welders by Chromium-Nickel Containing Additives." Staub—Reinhalt. Luft, 44(11):465–468 (1984)

(172) Zober, A. and Weltle, D. "Cross-Sectional Study of Respiratory Effects of Arc Welding." J. Soc. Occup. Med., 35(3):79–84 (1985)